ROBUSTNESS THEORY AND APPLICATION

ROBUSTNESS THEORY AND APPLICATION

BRENTON R. CLARKE
Murdoch University

WILEY

Registered Offices
John Wiley & Sons, Inc., 111 River Street, Hoboken, NJ 07030, USA

Editorial Office
111 River Street, Hoboken, NJ 07030, USA

For details of our global editorial offices, customer services, and more information about Wiley products visit us at www.wiley.com.

Wiley also publishes its books in a variety of electronic formats and by print-on-demand. Some content that appears in standard print versions of this book may not be available in other formats.

Limit of Liability/Disclaimer of Warranty
The publisher and the authors make no representations or warranties with respect to the accuracy or completeness of the contents of this work and specifically disclaim all warranties; including without limitation any implied warranties of fitness for a particular purpose. This work is sold with the understanding that the publisher is not engaged in rendering professional services. The advice and strategies contained herein may not be suitable for every situation. In view of on-going research, equipment modifications, changes in governmental regulations, and the constant flow of information relating to the use of experimental reagents, equipment, and devices, the reader is urged to review and evaluate the information provided in the package insert or instructions for each chemical, piece of equipment, reagent, or device for, among other things, any changes in the instructions or indication of usage and for added warnings and precautions. The fact that an organization or website is referred to in this work as a citation and/or potential source of further information does not mean that the author or the publisher endorses the information the organization or website may provide or recommendations it may make. Further, readers should be aware that websites listed in this work may have changed or disappeared between when this works was written and when it is read. No warranty may be created or extended by any promotional statements for this work. Neither the publisher nor the author shall be liable for any damages arising here from.

Library of Congress Cataloging-in-Publication Data

Names: Clarke, Brenton R., author
Title: Robustness theory and application / by Brenton R. Clarke.
Description: Hoboken, NJ : John Wiley & Sons, 2018. | Series: Wiley series in
    probability and statistics | Includes bibliographical references and index. |
Identifiers: LCCN 2018007658 (print) | LCCN 2018010629 (ebook) | ISBN
    9781118669501 (pdf) | ISBN 9781118669372 (epub) | ISBN 9781118669303 (cloth)
Subjects: LCSH: Robust statistics.
Classification: LCC QA276.A2 (ebook) | LCC QA276.A2 C53 2018 (print) | DDC 519.5–dc23
LC record available at https://lccn.loc.gov/2018007658

Cover Design: Wiley
Cover Image: © lvcandy/GettyImages

Set in 10/12pt Times by SPi Global, Pondicherry, India

Printed in the United States of America

V075375_060718

To my darling wife Erica and much loved sons Andrew and Stephen

CONTENTS

FOREWORD

It could be said that the genesis of this book came out of a unit which was on robust statistics and taught by Noel Cressie in 1976 at Flinders University of South Australia. Noel's materials for the lectures were gathered from Princeton University where he had just completed his PhD. Having been introduced to M-, L-, and R-estimators I shifted to the Australian National University in 1977 to work on the staff at the Statistics Department in the Faculties. There I enrolled part time in a PhD with Professor C. R. Heathcote (affectionately known as Chip by his colleagues and family) who happened to be researching the integrated squared error method of estimation and more generally the method of minimum distance. The common link between the areas of study, robust statistics and minimum distance estimation, was that of M- estimation. Some minimum distance estimation methods can be represented by M-estimators. A typical model used in the formulation of robustness studies was the "epsilon-contaminated normal distribution." In the spirit of John W. Tukey from Princeton University the relative performance of the estimator, usually of location, was to consider it in such contaminated models. It occurred to me that one could also estimate the proportion of contamination, epsilon, in such models and when I proposed this to Chip he became enthusiastic that I should work on these mixture models for estimation in my PhD. Chip was aware that the trend for PhDs was to have a motivating set of data and to this end he introduced me to recently acquired earthquake data recordings which could be modeled with mixture modeling. A portion of a large data set was passed on to me by Professor R.S. Anderssen (known as Bob), also at the Australian National University. Bob also introduced me to the Fortran Computing Language.

My brief was to compute minimum distance estimators on the earthquake data. In the mean time, Chip introduced me to Professor Frank Hampel's PhD thesis and several references on mixture modeling. After 1 year of trying to compute variance covariance matrices for the minimum distance estimation methods and for some reason failing to get positive definite matrices as was expected, I decided to come back to M-estimation and study the theory more closely. An idea germinated that I could study the M-estimator at a distribution other than at the model parametric family and other than at a symmetric contaminating distributions. This became the inspiration for my own PhD work.

I had the good fortune to then cross paths with Peter Hall. Chip who had been burdened with the duties as Dean of the Faculty of Economics at ANU took sabbatical at Berkeley for a year and Peter became my supervisor. Peter was always so cheerful and encouraging when it came to research. He was publishing a book on the "Martingale limit theory and its application" with Chris Heyde, and he encouraged me to read books and papers on limit theorems. I thus became interested in the calculus associated with limit theorems, and asymptotic theory of M-estimators. Chip returned to ANU in 1980 and kindly advised me on the presentation of my thesis and arranged for three quality referees, one of whom was Noel Cressie!

For some reason I wanted to go overseas and see the world. This was made possible with a postdoctoral research assistant position to study time series analysis at Royal Holloway College, University of London, in the period 1980–1982. While I worked on time series, I took my free time to put together my first major publication. Huber's (1981) monograph had come out. My paper was to illustrate that for a large class of statistics that could be represented by statistical functionals, which were in fact M-estimators, it was possible to inherit both weak continuity and Fréchet differentiability. These qualities in turn provide inherent robustness of the statistics. From the time of first submission to actual publication in *The Annals of Statistics* it took approximately 2.5 years to see it come out. It was during this time of waiting that I traveled to Zürich after writing to Professor Hampel. He was keen to see my work published as it supported with rigor notions which he had put forward in a heuristic manner, vis-à-vis the influence function. Subsequently, I spent almost a year at the Swiss Federal Institute of Technology (ETH), working as a research assistant and tutoring a class on the analysis of variance class lectured by Professor Hampel.

The Conditions A and discussion that are given in Chapter 2 of this book are from that Annals of Statistics paper. To facilitate the theory of weak continuity and Fréchet differentiability, I initially had to make smoothness assumptions on the defining ψ-functions for the M-estimators. It was not until I traveled to the University of North Carolina at Chapel Hill where I picked up the newly published book by Frank H. Clarke on "Optimization and Nons- mooth Analysis" that I realized how proofs of weak continuity and Fréchet differentiability for M-estimators with nonsmooth psi-functions, or psi-functions which were bounded and continuous but had "sharp corners," could follow through. I subsequently wrote a paper from

Murdoch University where I had taken up in 1984 a newly appointed lecturing position in the then Mathematics Department. The paper was eventually published in 1986 in *Probability Theory and Related Fields*. This book brings together both these papers and a paper on what are called selection functionals.

My sojourn at Murdoch University has been one of teaching and research. I benefited from many years of teaching in service course and undergraduate mathematics and statistics units, having developed materials for a unit on "Linear Models" which later became a Wiley publication in 2008. I have also developed a unit on "Time Series Analysis" and have two PhD students write theses in that general area. These forays while time consuming have helped me understand statistics a lot better. It has to be said that to teach robust statistics properly one needs to understand the mathematics that comes with it. Essentially, my experience in robust statistics has been one coming out of mathematics departments or at least statistics groups heavily supported by mathematics departments. But from the mathematics comes understanding and eventually new ideas on how to analyze data and further appreciation of why some things work and others do not. This book is a reflection of that general summary.

In writing this book I have also alluded to or summarized many works that have been collaborative. An author with whom I have published much is Pro- fessor Tadeusz Bednarski from Poland. I met Professor Bednarski at a Robust Statistics Meeting in Oberwolfach, Germany, in 1984. He recognized the importance of Fréchet differentiability and in particular the two works mentioned earlier and we proceeded to make a number of joint works on the asymptotics of estimators. He spoke on Fréchet differentiability at the Australian Statistics Conference 2010 held in Fremantle in Western Australia. However, with the tyranny of distance and our paths diverging since then it is clear that this book could not be written collaboratively. However, I owe much to the joint research that we did as is acknowledged in the book.

I have also benefited from collaborative works with many other authors. These works have helped in the presentation of new material in the book. In 1993 I published a paper with Robin Milne and Geoffrey Yeo in the *Scandinavian Journal of Statistics*. I thank both Robin and Geoff for making me think about the asymptotic theory when there are multiple solutions to ones estimat- ing equations. There are subsequently new examples and results on asymptotic properties of roots and tests in Chapter 4 of this book that have been developed by the author. In 2004 the author published a paper with former honors student Chris Milne in the Australian & New Zealand Journal of Statistics on a small sample bias correction to Huber's Proposal 2 estimator of location and scale and followed this with a paper in 2013 at the ISI meeting in Hong Kong. Summary results are included with permission in Section 5.1.2. In 2006 I collaborated with Andreas Futschik from the University of Vienna, to study the properties of the Newton Algorithm when dealing with either M-estimation or density estimation and a new Theorem 5.1 is borne out of that work. My interest in minimum distance estimation and its applications are summarized in Chapter 6. These include references to work with Chip Heathcote and also other collaborators such as Peter McKinnon and Geoff Riley. A new theorem on the unbiased nature of the vector parameter estimator of proportions given all other location and scale parameters in a mixture of normal distributions are known is given in Theorem 10.1. In addition plots in Figures 2.1, 6.5, 6.6, 6.7, 6.8, and 6.9 are reproduced with acknowledgment from their source.

No book on robustness is complete without the study of L-estimators or estimation of linear combinations of order statistics. I have only attempted to introduce the ideas which lead on to natural extensions on to least trimmed squares and generalizations to trimmed likelihood and adaptive trimmed likeli- hood algorithms. I have found these useful for identifying outliers where there are outliers to be found, yet caution the reader to use Fréchet differentiable es- timators for robust statistical inference. The outlier detection methods depart from the general use of Cooks distance in regression estimation yet have the appealing feature that they work even when there are what are termed points of influence.

The book does not canvas robust methods in time series or robust survival analysis, though references are given. Maronna et al. (2006) book is a useful starting point for robust time series. Developments on robust survival analysis continue to accrue. The presentation of this book is not exhaustive and many areas of endeavor in robust statistics are not countenanced in this book. The book mainly is a link between many areas of research that the author has been personally involved with in some way and attempts to weave the essence of relevant theory and application.

The work would never have been possible without the introduction of Fréchet differentiability into the statistical literature by Professor C. R. Rao and Professor G. Kallianpur in their ground-breaking paper in 1955. We have much to remember the French mathematician Maurice Renáe Fréchet for as well as Sir Ronald Aylmer Fisher who helped to motivate the 1955 paper.

PREFACE

This book requires a strong mathematics background as a prerequisite in or- der to be read in its entirety. Students and researchers will then appreciate the generally easily read Chapter 1 in the Introduction. Chapters 2 and 3 require a mathematics background, but it is possible to avoid the difficult proofs requiring the knowledge of the inverse function theorem in the proofs of weak continuity and Frèchet differentiability of M-functionals, should you need to gloss over the mathematics. On the other hand, great understanding can be gleaned from paying attention to such proofs. There are references later in the book to other important theorems such as the fixed point theorem and the implicit function theorem, though these are only referred to, and keen students may chase up their statements and proofs in related papers and mathematics texts. In this book Chapter 4 is important to the understanding that there can be more than one root to one's estimating equations and gives new results in this direction. Chapters 5–9 include applications and vary from the simple applications of computing robust estimators to the asymptotic theory descriptions which are a composite of exact calculation, such as in the theory of L_2 estimation of proportions, or descriptions of asymptotic normality results that can be further gleaned by studying the research papers cited. The attempt is to bring together works on estimation theory in robust estimation. I leave it to others to consider the potentially more difficult theory of testing, albeit robust confidence intervals based on asymptotics are a starting point for such.

This book has been written in what may be the last decade of my working career. Hopefully, others may benefit from the insights that this compendium of knowledge, which covers much research into robustness that I have had a part to play with, gives.

7 December 2017

BRENTON R. CLARKE
Perth, Western Australia

ACKNOWLEDGMENTS

I wish to acknowledge my two PhD supervisors Chip Heathcote and Professor Peter G. Hall. Both passed away in 2016. I remember them for their generous guidance in motivating my PhD studies in the period 1977–1980. Also I have to thank Professor Frank R. Hampel and his colleagues and students for helping me on my way during postdoctoral training as a Wissenschaftliche Mitarbeiter at ETH in 1982–1983. Their influence is unmistakeable. I owe much to the late Professor Alex Roberston for his help in bringing me to the then Mathematics Department, now called Mathematics and Statistics at Murdoch University, and I thank all my mathematics and statistics colleagues past and present for their generosity in allowing me to teach and research while at Murdoch University.

To my collaborators mentioned in the Foreword I give my thanks. Special thanks are to Professor Tadeusz Bednarski for fostering international collaboration in mathematics and statistics pre and post Communism (in Poland) and showing that there are no international borders religious or political in the common language of mathematics. I also thank Professor Andrzej Kozek who first hosted me at the University of Wroclaw in Poland and introduced me to Professor Bednarski.

Other researchers with whom I have the pleasure of being able to work with include Thomas Davidson, Andreas Futschik, David Gamble, Robert Hammarstrand, Toby Lewis, Peter McKinnon, Chris Milne, Robin Milne, Geoffrey Yeo, and Geoff Riley just to name some. More recent collaborations are with Christine Mueller and students. Robert Hammarstrand has also contributed by working under my direction to polish some of the R-algorithms associated with this book, for which I take full responsibility.

I have to acknowledge the work with Daniel Schubert. He wrote and gained his PhD under my supervision on the area of trimmed likelihood, but after gaining a position in CSIRO had his life cut short in a motor bike accident in 2007. I remember his eccentricities and for his enthusiasm for his newly found passion of robust statistics when he was a student. I include in the suite of associated R-algorithms our contribution to the Adaptive Trimmed Likelihood Algorithm for multivariate data.

As I came nearer the publication due date in July 2016, I had the privilege of visiting the Department of Statistics at The University of Glasgow headed by Professor Adrian Bowman. In August 2016 I visited Professor Mette Langaas and colleagues in the Statistics Group in the Mathematics Department at the Norwegian University of Science and Technology (NTNU). Some of this book was inspired by these visits. The journey was also facilitated by a visit to the University of Surrey, where I was hosted by Dr. Janet Godolphin in the Department of Mathematics. Finally, I acknowledge Murdoch University for the sabbatical that was taken nominally at the University of Western Australia, for the remainder of second semester 2016 used in preparation of this book. I thank Berwin Turlach at the University of Western Australia for his administrative role in arranging this. In addition I would like to thank Professor Luke Prendergast for encouragement and comments on a penultimate version of the book. Also thanks to Professors Alan Welsh and Andrew Wood for encouragement.

It goes without saying that I owe much to my wife and children in the formulation of this book and its predecessor *Linear Models: The Theory and Application of Analysis of Variance* which was published in 2008. They are duly acknowledged. Subsequently, I dedicate both books to them.

BRENTON R. CLARKE

NOTATION

$\in$	element of
$\notin$	not an element of
$A \cap B$	intersection of sets A and B
$A \cup B$	union of sets A and or B
$A \subset B$	A is contained in B
$\tilde{R}$	observation space
R	separable metrizeable space
$\mathcal{E}$	real line
$\mathcal{E}^+$	positive real line
$\mathcal{E}^r$	Euclidean r-space
$\mathcal{G}$	space of distribution functions
$\mathcal{A}^\delta$	closed delta neighborhood of set A
Θ	parameter space
$E[X]$	expected value of the random variable X
$\phi(z)$	standard normal density function
$\Phi(z)$	cumulative normal distribution
$\mathcal{I}(\theta)$	Fisher information
$|\cdot|$	modulus or absolute value of
$\rightarrow_p$	convergence in probability
$\rightarrow_d$	convergence in distribution
$\rightarrow_{a.s.}$	convergence almost surely
$\Rightarrow$	converges weakly
$\Longrightarrow$	implies

$\forall$	for every
$\|\cdot\|$	for vectors and matrices it is the Euclidean norm
	for elements on a normed linear space it is the norm
$N(\mu, \sigma^2)$	univariate normal random variable with mean μ and variance σ^2
$N(\mu, \Sigma)$	multivariate normal random variable
	with mean μ and covariance matrix Σ
$\Sigma_{i=1}^{n} x_i$	$x_1 + x_2 + \ldots + x_n$

ACRONYMS

ATLA   adaptive trimmed likelihood algorithm
CLT   central limit theorem
EM   expectation maximization
LLN   law of large numbers
MAD   median absolute deviation
MADN   normalized median absolute deviation
MCVM   minimum Cramèr–von Mises estimator
MLE   maximum likelihood estimator
SLLN   strong law of large numbers
TLE   trimmed likelihood estimator
WLLN   weak law of large numbers

ABOUT THE COMPANION WEBSITE

This book is accompanied by a companion website:

www.wiley.com/go/clarke/robustnesstheoryandapplication

BCS Ancillary Website contains:

- several computing programs, some in R and some in MatLab.
- The suite of routines were developed by Brenton and in some cases in conjunction with other students and collaborators of Brenton.
- No responsibility lies with Brenton or any others mentioned for the use of the routines given in endeavours outside the book.

BRENTON R. CLARKE
November 2017

1

INTRODUCTION TO ASYMPTOTIC CONVERGENCE

1.1 INTRODUCTION

The first and major proportion of this book is concerned with both asymptotic theory and empirical evidence for the convergence of estimators. The author has contributed here in more than just one article. Mostly, the relevant contributions have been in the field of M-estimators, and it is the purpose of this book to make some ideas that have necessarily been couched in deep theory of continuity and also differentiability of functions more easily accessible and understandable. We do this by illustration of convergence concepts which begin with the strong law of large numbers (SLLN) and then eventually make use of the central limit theorem (CLT) in its most basic form. These two results are central to providing a straightforward discussion of M-estimation theory and its applications. The aim is not to give the most general results on the theory of M-estimation nor the related L-estimation discussed in the latter half of the book, for these have been adequately displayed in other books including Jurečková et al. (2012), Maronna et al. (2006), Hampel et al. (1986), Huber and Ronchetti (2009), and, Staudte and Sheather (1990). Rather, motivation for the results of consistency and asymptotic normality of M-estimators is explained in a way which highlights what to do when one has more than one root to the estimating equations. We should not shy away from this problem since it tends to be a recurrent theme whenever models become quite complex having multiple parameters and consequently multiple simultaneous potentially nonlinear equations to solve.

1.2 PROBABILITY SPACES AND DISTRIBUTION FUNCTIONS

We begin with some basic terminology in order to set the scene for the study of convergence concepts which we then apply to the study of estimating equations and also loss functions. So we assume that there is an "Observation Space" denoted by $\tilde{R}$, which can be a subset of a separable metric space R. In published papers by the author, it was assumed typically that $\tilde{R}$ was a separable metric space, which did not allow for discrete data observed on, say, the nonnegative integers (such as Poisson distributed data). However, it is enough to consider now $\tilde{R} \subset R$ since the arguments follow through easily enough. So the generality of the discussion includes data that are either continuous or discrete, and either univariate or multivariate, or say defined on the positive real line such as in lifetime data.

For instance, if the data are k-dimensional continuous multivariate data, we have that $R \equiv \mathcal{E}^k$. Then we let $\mathcal{B}$ be the smallest σ-field containing the class of open sets on R generated by the metric on R. These are called the k-dimensional Borel sets, and $\mathcal{B}$ is known as the Borel σ-field. See Problem 1.1 for the definition of a σ-field.

A distribution on R is a nonnegative and countably additive set function, μ, on $\mathcal{B}$, for which $\mu\{R\} = 1$, and it is well known say that on $R = \mathcal{E}$ there corresponds a unique right continuous function F whose limits are zero and one at $-\infty$ and $+\infty$, defined by $F(x) = \mu\{(-\infty, x]\}$.

We shall denote an abstract probability space to be $(\Omega, \tilde{\mathcal{A}}, \mathcal{P})$. This formulation assumes $\tilde{\mathcal{A}}$ to be a σ-field of subsets of Ω, with $\mathcal{P}$ a probability measure on $\tilde{\mathcal{A}}$. Ω is thought of as a sample space and elements of Ω, denoted by ω, are the outcomes. Then a sequence of random variables on Ω is defined via

$$X(\omega) = X_1(\omega), X_2(\omega), \ldots, X_n(\omega), \ldots, \quad (1.1)$$

taking values in the infinite product space $(R^\infty, \mathcal{B}^\infty)$. The observed sample of size n is then written as

$$(X_1(\omega), \ldots, X_n(\omega)) = \pi^{(n)} o X(\omega),$$

while the nth random variable is given by $X_n(\omega) = \pi_n o X(\omega)$. Both $\pi^{(n)}$ and π_n are then what are termed measurable maps with respect to $\mathcal{B}^\infty$. They induce distributions $G^{(n)}$ and G_n on $(R^n, \mathcal{B}^n)$ and $(R, \mathcal{B})$, respectively. A useful reference on equivalent representations of infinite sequences of random variables and probability measures is that of Chung (2001), (now 3rd ed.) (1st ed., pp. 54–58).

We use the symbol $\mathcal{G}$ to denote the space of distributions on $(R, \mathcal{B})$. Consider an arbitrary set $A^{(n)} = (A_1 \times A_2 \times \ldots \times A_n) \in \mathcal{B}^n$. Denote by $G^{(n)}$ the product measure on $(R^n, \mathcal{B}^n)$ that gives

$$P(\pi^{(n)} X(\omega) \in A_1 \times A_2 \times \ldots \times A_n) = \int_{A^{(n)}} dG^{(n)}.$$

Then we say that the sequence X is independent identically distributed (*i.i.d.*) if there exists a $G \in \mathcal{G}$ such that for every $A^{(n)}$ in the form above

$$\int_{A^{(n)}} dG^{(n)} = \int_{A_1} dG \times \ldots \times \int_{A_n} dG.$$

1.3 LAWS OF LARGE NUMBERS

The law of large numbers (LLN) is a result that describes what will happen if we repeat an experiment a large number of times. To couch it in simple terms, it is when the average of the results/experiments obtained from a large number of trials should be close to its expected value. How we measure closeness and relate that to a limit of the number of trials, tending to infinity say, requires us to introduce two possible modes of convergence of random variables.

1.3.1 Convergence in Probability and Almost Sure

For a sequence $\{X_n\}$ of random variables on $(\Omega, \tilde{\mathcal{A}}, P)$, we can define two modes of convergence to a random variable X on that same probability space. The first and weaker mode of convergence, convergence in probability, is defined as follows: Let d be the metric distance on R. Then we say, provided the random variables are defined on the same probability space, X_n converges in probability to X if for every $\epsilon > 0$

$$P\{\omega \in \Omega | d(X_n, X) < \epsilon\} \to 1 \quad \text{as } n \to \infty. \tag{1.2}$$

We say

$$X_n \to_p X. \tag{1.3}$$

Should X be a constant and say $\tilde{R} \subset \mathcal{E}^k$, then say $P(X = \text{const}) = 1$ the definition can be modified to say $X_n \to_p \text{const}$ if for every $\epsilon > 0$

$$P\{|X_n - \text{const}| < \epsilon\} \to 1 \quad \text{as } n \to \infty.$$

A stronger form of convergence of say X_n to the random variable X is convergence with probability one or almost sure convergence

$$P\{\omega : \lim_{n\to\infty} X_n(\omega) = X(\omega)\} = 1.$$

It is a fact that this form of convergence implies convergence in probability. Hence, statements of almost sure convergence can be preferred to statements made only in probability.

1.3.2   Expectation and Variance

In order to relate the LLN, we need a formal concept of the expected value of a random variable. For a random variable on $(\Omega, \tilde{A}, \mathcal{P})$, its expected value is written $E[X] = \int_\Omega X(\omega)dP(\omega) = \int_R xdG(x)$, where G is the induced distribution function on the full space R that contains the observation space $\tilde{R}$. Thus, for observations on a subset of the real line, that is where $R \subset \mathcal{E}$ we have then, denoting $\mu = E[X]$,

$$\mu = \int_{-\infty}^{+\infty} xdG(x).$$

This is the mean value of the random variable X. Similarly, we may define the variance assuming it exists, via

$$\text{var}[X] = E[(X - \mu)^2] = \int_\Omega (X(\omega) - \mu)^2 dP(\omega) = \int_R (x - \mu)^2 dG(x).$$

As is typically done, we denote the variance $\sigma^2 = \text{var}(X)$. The variance measures the spread of the observations about the expected value. Since the variance is involving squared deviations from the mean, it is not in the same units of measure as the original variable. Hence, what is more often used as a measure of spread is the standard deviation which is the square root of the variance, that is $\sigma = \sqrt{\sigma^2}$. Nevertheless, the variance plays an important part in convergence concepts and indeed makes proofs, even of the LLN in its weaker form, much easier when it is known to exist, that is when it is finite.

1.3.3   Statements of the Law of Large Numbers

The LLN in either its weaker or stronger form state that – with almost certainty - the sample average

$$\overline{X}_n = \frac{1}{n}(X_1 + X_2 + \ldots + X_n)$$

converges to the expected value (assumed finite) so that

$$\overline{X}_n \to \mu \quad \text{for } n \to \infty.$$

Here $X_1, X_2, \ldots$ is assumed *i.i.d.* and consequently $E[X_1] = \ldots = E[X_n] = \mu$. It is not necessary that $\mathrm{var}[X_1] = \ldots = \mathrm{var}[X_n] = \sigma^2 < +\infty$. The LLN will hold anyway. Large or infinite variance makes convergence "slower." Finite variance makes proofs easier and shorter. Other ways that the theorem can be extended is by relaxing assumptions of random variables being independent and/or identically distributed. Such considerations may become important when allowing for the incorporation of ideas such as often promulgated in the robustness literature, where it is usually the case that model assumptions can and often do fail. These ideas will become more apparent once we travel on to statistical inference.

Statement of the Weak Law of Large Numbers

The weak law of large numbers (WLLN) states that

$$\overline{X}_n \to_p \mu \quad \text{as} \quad n \to \infty$$

Equivalently, for any $\epsilon > 0$

$$\lim_{n \to +\infty} P\{|\overline{X}_n - \mu| > \epsilon\} = 0.$$

Interpretation of the WLLN is that for any, however, small nonzero deviation from the mean (governed by setting a value for ϵ say), with a suitably large sample size, there is a high probability that the sample average is close to the expected value, that is within that deviation from the mean.

The WLLN involves convergence in probability and there are examples where a random variable converges in probability but does not converge almost surely. Hence, this is the reason for it to be the weak LLN.

Statement of the SLLN

The SLLN states that

$$P\{\omega | \lim_{n \to +\infty} \overline{X}_n(\omega) = \mu\} = 1$$

or put more succinctly,

$$\overline{X}_n \to_{a.s.} \mu.$$

As noted above, convergence almost surely implies convergence in probability, whence the SLLN implies the WLLN.

1.3.4 Some History and an Example

Probability theory, both modern and historical, has been inspired by the need to analyze games of chance. See for instance that the Italian mathematician Geralamo Cardono, who was well versed in games of chance, is credited by Mlodinow (2008) with writing down a version of the law of large numbers, stating without proof that "the accuracies of empirical statistics tend to improve with the number of trials." Indeed, Jakob Bernoulli (1713) examined a special form of the LLN

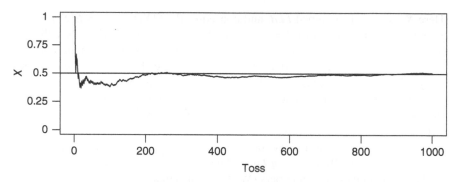

FIGURE 1.1   Illustration of the SLLN in action by plotting the average of Bernoulli random variables with probability $p = 0.5$ as $n =$ number of tosses becomes large.

for a binary random variable and is credited with the first proof. To illustrate the LLN in practice, we consider a sequence of Bernoulli random variables, where with probability one half the outcome of a random variable in the sequence is one, whereas with probability one half the outcome of the random variable is zero. For instance, we may be watching a sequence of independent coin tosses where the random variable is denoted a one if a head occurs or a zero if a tail occurs. Constructing such a sequence is known to be fraught with difficulty as can be gained from reading Diaconis et al. (2007). Of course Bernoulli random variables can be assigned more generally if we had "biased" coins where the probability of a head was some p where $0 < p < 1$, not necessarily $p = \frac{1}{2}$, and the consequent probability of a tail is $q = 1 - p$. In terms of a sample space and possible sequences of random variables, we can consider a random number generator on a computer starting an infinite sequence, the random seed to start that sequence being generated by the time of day. Assuming an infinite number of possible starting points if we choose just one starting point, that is a particular $\omega \in \Omega$ with probability one, we will have chosen a sequence for which the average of those observed coin tosses converges to the expected value of the random variable. Basic maths can show that the expected value of the Bernoulli random variable is a weighted average of the outcomes, $\mu = 1 \times p + 0 \times (1 - p) = p$, whence in the limit if we set $p = 1/2$ if we observe, with probability one, just one sequence for long enough we should see the average of the random variable converges to 0.5. A randomly selected graph is given in Figure 1.1, which demonstrates the convergence by plotting the sample average for up to one thousand "tosses" of the Bernoulli random number generator.

1.3.5   Some More Asymptotic Theory and Application

Consider now a sequence of *i.i.d.* random variables $X_1, X_2, \ldots$. For clarity of exposition, we give the following formulation that helps to relate statements about events to almost sure convergence. *Definition 1* The sequence of statements $A_1(X_1), A_2(X_1, X_2)$, ... are said to hold *for all sufficiently large n*, (*f.a.s.l.n*) if

$$P\{\omega|\ \cup_{m=1}^{\infty} \cap_{n=m}^{\infty} A_n(X_1(\omega),\ \dots\ , X_n(\omega))\} = 1.$$

Using this definition it is clear that a sequence of measurable maps $\{T_n(X_1,\ \dots\ , X_n)\}$ from the domain space R^n to the codomain $\mathcal{E}^r$ converges almost surely to T if and only if for every $\eta > 0$ the sequence of statements

$$|T_n(X_1,\ \dots\ , X_n) - T| < \eta \quad n = 1, 2,\ \dots$$

holds $f.a.s.l.n$. For an account of the above definition, see Foutz and Srivastava (1979). Interestingly, we have that if we have two sequences of statements that are both known to hold $f.a.s.l.n$ say

$$A_1(X_1), A_2(X_1, X_2),\ \dots$$

and

$$B_1(X_1), B_2(X_1, X_2),\ \dots$$

then the sequence of statements

$$A_1(X_1) \cap B_1(X_1), A_2(X_1, X_2) \cap B_2(X_1, X_2),\ \dots$$

also holds $f.a.s.l.n$. This is for the simple reason that since both

$$A = \cup_{m=1}^{\infty} \cap_{n=m}^{\infty} A_n(X_1(\omega),\ \dots\ , X_m(\omega))$$

and

$$B = \cup_{m=1}^{\infty} \cap_{n=m}^{\infty} B_n(X_1(\omega),\ \dots\ , X_m(\omega))$$

are both sets of probability one whence $P\{A \cap B\} = 1$, and since also we have the identity

$$A \cap B = \{\cup_{m=1}^{\infty} \cap_{n=m}^{\infty} A_n\} \cap \{\cup_{m=1}^{\infty} \cap_{n=m}^{\infty} B_n\} = \cup_{m=1}^{\infty} \cap_{n=m}^{\infty} (A_n \cap B_n)$$

The implications of this result are that to check any finite number of statements that hold $f.a.s.l.n$. they then hold simultaneously together $f.a.s.l.n$. Thus, automatically if say one set of statements is that there is a root of a random function in say an interval $(\mu - \eta, \mu + \eta)$ and also it is known that the random function is monotonic in the same interval and both statements hold $f.a.s.l.n$, then this allows us to conclude there is a unique zero in that interval $f.a.s.l.n$, for example.

Returning to the SLLN, if there is any real valued measurable function $f(\cdot)$ from the space R to the space $\mathcal{E}$, and as before $X_1, X_2,\ \dots$ are *i.i.d.* variables, then

$f(X_1), f(X_2), \ldots$, forms an independent sequence of random variables on $\mathcal{E}$. If in addition $E[|f(X)|] < +\infty$, then by the SLLN

$$\frac{f(X_1) + f(X_2) + \ldots + f(X_n)}{n} - E[f(X)] \to_{a.s.} 0. \tag{1.4}$$

The property (1.4) is known to hold for more general sequences than *i.i.d.* sequences and is known as an "ergodic property." Ergodic sequences include stationary sequences satisfying appropriate mixing conditions, but suffice to say the discussion in this book relates in the main to *i.i.d.* random variables.

Now it is possible that one may be interested in setting up a class of functions $\mathcal{A}$ which could be infinite in number, yet one can achieve the following result

$$\sup_{f \in \mathcal{A}} |\frac{f(X_1) + f(X_2) + \ldots + f(X_n)}{n} - E[f(X)]| \to_{a.s.} 0. \tag{1.5}$$

Here at the very least we may require for each $f \in \mathcal{A}$ to be bounded in modulus above by some g where $E[|g(X)|] < +\infty$. In addition, we may also require the functions f to satisfy certain equicontinuity or maybe even stronger continuity conditions which we shall elaborate later. For instance as in Rao (1962), $\mathcal{A}$ is equicontinuous at each $x \in X$, i.e. for each $\epsilon > 0$ there exists a neighborhood N of x such that $|f(y) - f(x)| < \epsilon$, for all $y \in N$ and all $f \in \mathcal{A}$.

1.4   THE MODUS OPERANDI RELATED BY LOCATION ESTIMATION

To illustrate why we would concern ourselves with these abstract concepts, we shall examine the class of functions associated with what is known as the Tukey bisquare function. This is governed by the formula

$$\psi_{BS}(x) = \begin{cases} x(1 - x^2/c^2)^2 & -c < x < c \\ 0 & \text{Otherwise} \end{cases} \tag{1.6}$$

The constant $c > 0$ may be chosen to be fixed, and for illustration purposes we choose it to be the value 4.685. See Figure 1.2. This function, which is an odd function, is used in many discussions of what are called M-estimators. It was first introduced by Beaton and Tukey (1974) and is prominent in the discussion of M-estimators of location parameters, particularly when it is possible that one has potential contamination in the tails of one's data set. To set the scene, we can imagine a normal parametric location model, that is our sequence of data is generated from a normal distribution with mean μ and the model parametric family is

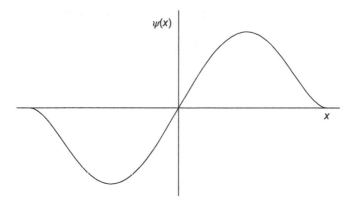

FIGURE 1.2   Illustration of the Tukey bisquare psi function.

$\{F_\tau(x) = \Phi(x - \tau) : \tau \in \mathcal{E}\}$, where $\Phi(x) = \int_{-\infty}^{x} \frac{1}{\sqrt{2\pi}} \exp(-u^2/2) du$. To estimate the parameter μ given our sequence of data, we solve the equations

$$K_n(\tau) = \frac{1}{n} \sum_{i=1}^{n} \psi(X_i - \tau) = 0 \qquad (1.7)$$

For any fixed τ the normalized sum converges by the SLLN to $K_G(\tau) = \int \psi(x - \tau)\phi(x - \mu)dx$, where $G = F_\mu$ and $\phi(x) = \Phi'(x)$ is the normal density. The equations are such that $K_G(\mu) = 0$, hence a good candidate for an estimate of μ is a zero or root of the empirical equations (1.7). Our class of functions then is $\mathcal{A} = \{\psi(\cdot - \tau) : \tau \in \mathcal{E}\}$. Without loss of generality we set $\mu = 0$ and examine the empirical functions $K_n(\tau)$ that you see for instance in generating several data sets. See for instance Figure 1.3. For such a small sample size, it is clear that the SLLN is only just beginning to work. The 10 plots exhibit a great deal of variability; however, it can be noticed that the zeros of the functions $K_n(\tau)$ that are in the middle of the graphs are centered approximately around the true mean of $\mu = 0$. Clearly also there are infinitely many zeros in the tails of these redescending curves, which carry little or no information about the central roots of the curves. They are an artifact of the redescending nature of ψ_{BS}. Increasing the sample size to $n = 50$ in Figure 1.4 shows the fairly uniform convergence of the empirical curves $K_n(\tau)$ to the asymptotic curve, in this case $K_\Phi(\tau)$, and this convergence is even more graphic in the plots of Figures 1.5 and 1.6 when $n = 100$ and 500, respectively.

Interestingly, the roots about the central value of zero are asymptotically unique. To support this claim, we can examine the derivative of the function $K_n(\tau)$, call it $K_n'(\tau)$. See Figure 1.7. For a sample size of $n = 100$, it is clear that for $\tau \in [-1, 1]$ the function derivative has a value which is almost certainly less

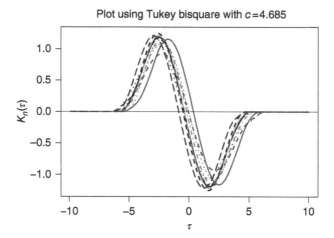

FIGURE 1.3 Illustration of SLLN for the family of functions involving the Tukey bisquare; sample size $n = 5$. Plots are for 10 independent samples.

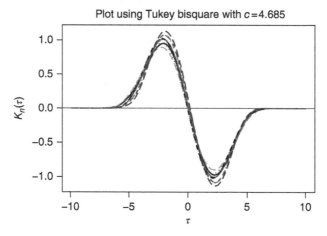

FIGURE 1.4 Illustration of SLLN for the family of functions involving the Tukey bisquare; sample size $n = 50$. Plots are for 10 independent samples.

than say -0.2. For large n it is also clear that the zero of $K_n(\tau)$ is say inside $[-1, 1]$. We therefore conclude it is a unique zero in this region since the empirical curve is almost certainly monotonically decreasing.

The appearance of the Tukey bisquare function came after the introduction of redescending M-estimators of location, initially proposed in Andrews et al. (1972) and further related in Hampel (1974). Due to the existence of multiple roots of the estimating equations, Hampel's recommendation was to choose the root closest to the median. Since the median is consistent to the true $\mu = 0$, it is therefore clear that the root that is closest to the median should be the asymptotically unique

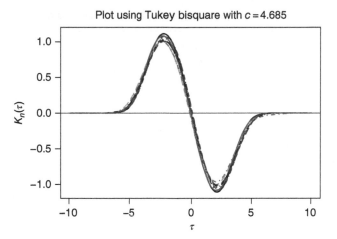

FIGURE 1.5 Illustration of SLLN for the family of functions involving the Tukey bisquare; sample size $n = 100$. Plots are for 10 independent samples.

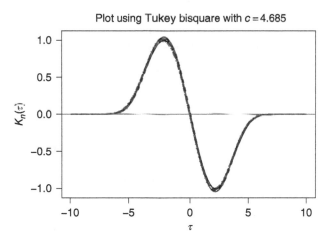

FIGURE 1.6 Illustration of SLLN for the family of functions involving the Tukey bisquare; sample size $n = 500$. Plots are for 10 independent samples.

consistent root of the estimating equations, given our discussion above. Of course numerically finding a unique consistent root of the equations relies on the choice of algorithm. By far the most well-known algorithm is that of Newton–Raphson, where the iterations are

$$\tau_{v+1}^{(n)} = \tau_v^{(n)} - \frac{K_n(\tau_v^{(n)})}{K_n'(\tau_v^{(n)})}, \quad v = 0, 1, 2, \ldots \tag{1.8}$$

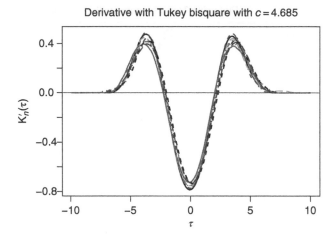

FIGURE 1.7   Illustration of SLLN for the family of functions involving the derivative of the Tukey bisquare; sample size $n = 100$. Plots are for 10 independent samples.

and where the superscript (n) indicates dependence on the random equations (1.7). Convergence of the iteration to a unique consistent root ξ_n, starting with a suitable initial, possibly random, estimate, relies on almost sure uniform convergence of the sequence $K_n(\tau)$ and its derivative $K'_n(\tau)$ to $K_G(\tau)$ and $K'_G(\tau)$, respectively. That is, for any given $\delta > 0$ and arbitrary compact sets C the sequence of statements

$$\sup_{\tau \in C} |K_n(\tau) - K_G(\tau)| < \delta, \quad \sup_{\tau \in C} |K'_n(\tau) - K'_G(\tau)| < \delta, \quad n = 1, 2, \ldots \tag{1.9}$$

hold $f.a.s.l.n$. Here sup denotes the least upper bound (or supremum) of the set. These conditions are easily derived in the case of the Tukey bisquare, ψ_{BS}, as it is a bounded smooth function with continuous and bounded derivatives. See the discussion and proof of Theorem 1 and Lemma 5.1 in Clarke (1986a). Note that the convergence given there is shown to be uniform over the whole real line, replacing C by $\mathcal{E}$. Thus, the figures above reflect this asymptotic convergence seen in Eq. (1.9).

Indeed the paper of Clarke (1986a) goes on to discuss the region from which the Newton–Raphson iteration converges to the central root of the estimating equations. Such a region is asymptotically governed by the curve $K_G(\tau)$ and so an upper limit on such a region is a solution μ^* of the equation

$$S_G(\tau) = 2(\tau - \mu) - \frac{K_G(\tau)}{K'_G(\tau)} = 0.$$

This equation gives the boundary values for the Newton–Raphson domain of attraction in the asymptotic equation $K_G(\tau) = 0$ with respect to the root μ. Without losing generality setting $\mu = 0$ the smallest positive root when $G = \Phi$

is $\mu_m^*=1.5737$. Starting from anywhere in the interval $(-1.5737, 1.5737)$, the Newton–Raphson algorithm applied to the asymptotic equation $K_\Phi(\tau) = 0$ will converge to $\mu = 0$. Obviously, for a finite sample there will be adjustments to make, which are detailed in that paper, but starting from a consistent estimate of μ, such as the median, one can see that the Newton–Raphson algorithm will converge to the root of interest.

It remains then to describe the properties of the root of interest, especially since one not only wishes for a point estimate but rather an idea of the variability of the point estimate. To explain this, we introduce the Lindeberg–Lévy CLT. Consider a sequence of random variables which are independent and identically distributed with mean μ and variance $\sigma^2 < +\infty$ and suppose we are interested in the sample average

$$\overline{X}_n = \frac{X_1 + \ldots + X_n}{n}.$$

The law of large numbers gives the asymptotic almost sure limit for the average as being μ. The classical CLT describes the way the averages fluctuate about the asymptotic limit of μ as they converge.

Theorem 1.1: *Lindeberg–Lévy CLT. Suppose $\{X_1, X_2, \ldots\}$ is a sequence of independent identically distributed random variables with $E[X_i] = \mu$ and $\mathrm{var}[X_i] = \sigma^2 < +\infty$. Then as* n *approaches* ∞, *the random variable* $\sqrt{n}(\overline{X}_n - \mu)$ *converges in distribution to a $N(0, \sigma^2)$ random variable.*

$$\sqrt{n}(\overline{X}_n - \mu) \to_d N(0, \sigma^2).$$

For $0 < \sigma < +\infty$, convergence in distribution means that the cumulative distribution functions of $\sqrt{n}(\overline{X}_n - \mu)$ converge pointwise in the sense that

$$\lim_{n \to +\infty} P\{\sqrt{n}\left(\overline{X}_n - \mu\right) \le z\} = \Phi\left(\frac{z}{\sigma}\right)$$

The convergence is uniform in z where

$$\lim_{n \to +\infty} \sup_{z \in \mathcal{E}} |P\{\sqrt{n}(\overline{X}_n - \mu) \le z\} - \Phi\left(\frac{z}{\sigma}\right)| = 0$$

It is now possible to apply this result to the central root of the estimating equations (1.7). Visualize or denote $\{\hat{\mu}_n\}$ to be the central root of the equations, for example as obtained by a Newton–Raphson iteration limit starting from say a median. To be explicit, we give the existence of an asymptotically unique consistent estimator here, since the proof illustrates the sort of continuity arguments in play and which can be extended to multivariate parameters in Chapter 2. Since not all ψ-functions for M-estimators are continuously differentiable (even though

ψ_{BS} is) we will relax the conditions accordingly and adapt the notation to this chapter.

Lemma 1.1: *(Clarke, 1986a, Lemma 8.1) Assume ψ is continuous. Denote by $K'_n(\tau)$ a left-hand derivative of K_n and suppose for any compact C Eq. (1.9) holds f.a.s.l.n. Let $K_G(\mu)$ be continuously differentiable, $K_G(\mu)=0$ and $K'_G(\mu) < 0$. Then there exists an interval of some radius $\delta > 0$ about μ, such that*

   (a) *there exists a consistent sequence $\{\mu_n\}$ of zeros of $\{K_n(\tau)\}$ in $(\mu - \delta, \mu + \delta)$, and*

   (b) *for any other sequence $\{\tilde{\theta}_n\}$ of zeros of $\{K_n(\tau)\}$ in $(\mu - \delta, \mu + \delta)$, $\tilde{\theta} = \mu_n$ f.a.s.l.n.*

   Proof. By continuity let $\delta > 0$ so that $K'_G(\tau) < \frac{1}{2}K'_G(\mu) = y_0$ say, for $\tau \in (\mu - \delta, \mu + \delta)$. From Eq. (1.9),

$$K'_n(\tau) < \frac{1}{2}y_0 \text{ uniformly in } \tau \in [\mu - \delta, \mu + \delta] \tag{1.10}$$

Then by Eq. (1.9) and the mean value theorem the statement

$$K_n(\mu - \delta) > K_G(\mu - \delta) + y_0\delta > 0 = K_G(\mu) > K_G(\mu + \delta) - y_0\delta > K_n(\mu + \delta)$$

holds f.a.s.l.n. Continuity of $K_n(\tau)$ implies existence of a zero $\mu_n \in (\mu - \delta, \mu + \delta)$ f.a.s.l.n. and by Eq. (1.10) this zero must be unique in the given interval. Since δ is arbitrary $\mu_n \to_{a.s.} \mu$.   □

Having established the existence of an asymptotically unique consistent sequence of roots, we find from the mean value theorem

$$0 = K_n(\hat{\mu}_n) = K_n(\mu) + K'_n(\xi_n)(\hat{\mu}_n - \mu), \tag{1.11}$$

where ξ_n is a point lying somewhere between $\hat{\mu}_n$ and μ. Recalling Eq. (1.9) and the facts that $K'_G(\cdot)$ is continuous and also necessarily $\xi_n \to_{a.s.} \mu$, it is clear that $K'_n(\xi_n) \to_{a.s.} K'_G(\mu) = \int_{-\infty}^{+\infty} \psi'_{BS}(x)\phi(x)dx$, which also implies

$$K'_n(\xi_n) \to_p K'_G(\mu) = -\int_{-\infty}^{+\infty} \psi'_{BS}(x)\phi(x)dx.$$

We now intend to invoke Slutsky's theorem. Consider two sequences $\{U_n\}$ and $\{V_n\}$, where U_n converges in distribution to a random variable U and V_n converges in probability to a constant k. Then $U_n/V_n \to_d U/k$ provided k is invertible. Rewriting Eq. (1.11) we have

$$\sqrt{n}(\hat{\mu}_n - \mu) = \frac{-\sqrt{n}K_n(\mu)}{K'_n(\xi_n)}. \tag{1.12}$$

It is now easy to see in virtue of the CLT that in the case where $G = \Phi$, that is the normal model holds, then

$$\sqrt{n}K_n(\mu) = \frac{1}{\sqrt{n}} \sum \psi_{BS}(X_i - \mu) \tag{1.13}$$

$$\to_d N\left(0, \int_{-\infty}^{+\infty} \psi_{BS}^2(x)\phi(x)dx\right) \tag{1.14}$$

here noting that the sequence of variables $\{\psi_{BS}(X_i - \mu)\}_{i=1}^{\infty}$ are independent with zero mean (the curve $\psi_{BS}(\cdot)$ is odd) and variance $E[\psi_{BS}^2(Z)]$, where Z is a standard normal random variable. Now employing Slutsky's theorem we find

$$\sqrt{n}(\hat{\mu}_n - \mu) \to_d N\left(0, \sigma_{BS}^2 = \frac{E[\psi_{BS}^2(Z)]}{E[\psi'_{BS}(Z)]^2}\right).$$

As a reflection of this asymptotic result, and what it may tell us about the location of the true mean, we refer to Figure 1.8 which is a histogram of 10 000 location estimates, found by starting at the median and doing five iterations of the Newton–Raphson algorithm. These are obtained using the Tukey bisquare as detailed earlier with sample size $n = 100$. Notice that the histogram is bell shaped, reflecting asymptotic normality of the estimates. Another way of viewing these results is

$$P\{\sqrt{n}\left(\frac{\hat{\mu}_n - \mu}{\sigma_{BS}}\right) \le z\} \to \Phi(z)$$

which suggests that a 95% confidence interval for the true mean given an estimate from any one sample is $(\hat{\mu}_n - 1.96 \times \sigma_{BS}/\sqrt{n}, \hat{\mu}_n + 1.96 \times \sigma_{BS}/\sqrt{n})$. In this example with $c = 4.685$, the value of $\sigma_{BS} = 1.026$ whereupon approximately 95% of estimates from the simulation will lie between ± 0.201 from the true mean which is zero in the illustration.

The primary aim of this section has been to introduce ideas of asymptotic convergence that can be illustrated, in this instance, using the vehicle of the redescending Tukey bisquare function, heavily used in the theory of robustness. As yet we

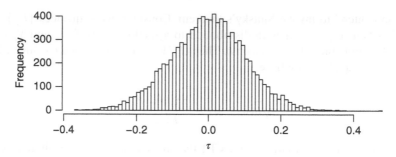

FIGURE 1.8   Histogram of the Tukey bisquare psi function estimates.

have not delved into what is a robust estimator? This is the subject of the next subsection and next chapter. The above analysis makes use of the uniform convergence over a class of functions, essentially gained through the action of the SLLN at each parameter value.

One of the first and very important results involving uniform convergence is called the Glivenko–Cantelli theorem, which asserts that

$$P\{\omega|\lim_{n\to\infty}\ \sup_{-\infty<x<+\infty}\ |F_n(x,\omega)-G(x)|=0\}=1. \qquad (1.15)$$

Here $F_n(x,\omega)$ is the *empirical distribution function* which is the distribution (random) that assigns atomic mass $1/n$ to each point of the sample $\pi^{(n)}oX(\omega)$. The proof of Eq. (1.15) stems from the SLLN applied to the sequence of indicator function values $\{I_{(-\infty,x]}(X_i)\}_{i=1}^{\infty}$. Extensions of the Glivenko–Cantelli theorem to $\mathcal{E}^k$ space and more general sets have been carried out by Wolfowitz (1954), Rao (1962), Topsoe (1970), and Elker et al. (1979). All of these consider the theorem for closed half spaces in $\mathcal{E}^k$. A more general theory is available in the work of Vapnik and Chervonenkis (2004) which also has applications in machine learning but suffice to say so long as your functions are smooth in some sense as in the form of the Tukey bisquare we do not need the abstractness of the most general theory. See also Pollard (1990) for further results on empirical processes.

Interestingly, the estimator $\hat{\mu}_n$ which is a solution to Eq. (1.7) can be thought of as an implicitly defined estimator that is a function of the empirical distribution function. For example,

$$K_n(\tau)=\int_{-\infty}^{+\infty}\psi(x-\tau)\mathrm{d}F_n(x),$$

so that we can think of $K_n(\tau)\equiv K_{F_n}(\tau)$. Defining the median to be $F_n^{-1}(1/2)=\inf\{u|F_n(u)\geq 1/2\}$ and the functional solution such that if we consider the set of solutions

$$S(\psi_{\mathrm{BS}},F_n)=\{\tau|K_n(\tau)=0\} \qquad (1.16)$$

then the estimator can be thought of as a function of the empirical distribution function by letting $\rho(F_n, \tau) = |\tau - F_n^{-1}(1/2)|$ and defining the estimator via

$$\inf_{\tau \in S(\psi_{BS}, F_n)} \rho(F_n, \tau) = \rho(F_n, T[F_n]). \qquad (1.17)$$

The estimator $T[F_n]$ is then both a function of the empirical distribution function and the function ψ_{BS} and the "selection functional" ρ. The idea of a selection functional was first coined in Clarke (1983a) and elaborated on in Clarke (1990). Then $T[F_n] \equiv T[\psi_{BS}, \rho, F_n]$. In the unlikely event that there are two solutions to Eq. (1.17) minimizing the distance from the median, one can choose the one on the left, effectively introducing a third tier of selection, though this is generally not needed. The functional ρ need not be the same as the objective functional used to derive the estimating equations, though this does not exclude such functionals.

In summary we have seen an explicit example illustrating the action of the SLLN. We have quoted from the literature the Glivenko–Cantelli result which yields one avenue regarding the convergence of the empirical distribution, F_n to the underlying distribution which can be any $G \in \mathcal{G}$. For a specific example, we consider corresponding uniform convergence of $K_{F_n}(\tau)$ to $K_\Phi(\tau)$ and also $K'_{F_n}(\tau)$ to $K'_\Phi(\tau)$. For the central asymptotic unique consistent root that results, we find that so long as $K'_\Phi(\mu) \neq 0$, the resulting central root behaves asymptotically normal, with a well-defined asymptotic variance. None of these results are new, but they prove an interesting backdrop to the results that shall later come and demonstrate clearly the connection between empirical convergence concepts and the behavior of some useful estimators. If we were to retain the asymptotic distribution as being normal, the efficient estimator $\overline{X}$ would narrowly beat the estimator that results from employing the Tukey bisquare with tuning constant $c = 4.685$, the asymptotic variance ratio being 0.95. However, as discussed graphically in the works by Tukey (1960) and Huber (1964), the dominance of the classical estimates at the model is soon overshadowed if we consider underlying distributions in ϵ contaminated distributions.

1.5   EFFICIENCY OF LOCATION ESTIMATORS

Based on asymptotic efficiency Tukey (1960) compared the estimator $\overline{X}$ with trimmed means and the median at distributions of the form $G(x) = (1 - \epsilon)\Phi(x) + \epsilon\Phi(x/3)$. Here Tukey is intending that ϵ is the average proportion of contaminated data entering the sample with an increased variance of nine. The idea of asymptotic efficiency assumes that the best one can do at a parametric distribution is written down in terms of the inverse of Fisher information. This is encapsulated in the Cramér–Rao inequality. For instance, let $T = T(\pi^{(n)} oX(\omega))$ be an unbiased estimator of μ, meaning $E[T] = \mu$. Here $\pi^{(n)} oX(\omega)$ has a probability function $f_\mu(\pi^{(n)}oX)$. Notably, the support of f does not depend on μ and under mild conditions of regularity which hold here.

$$\text{var}(T) \geq \cfrac{1}{E\left\{\left(\frac{\partial}{\partial\mu}\log f_\mu(\pi^{(n)}oX)\right)^2\right\}} \tag{1.18}$$

The denominator in Eq. (1.18) is denoted the *Fisher information* in $\pi^{(n)}oX$ about μ. Thus, if $\pi^{(n)}oX$ is a random sample from $f_\mu(x)$, the information in $\pi^{(n)}oX$ is n times the information in a single observation:

$$\mathcal{I}(\mu|\pi^{(n)}oX) = n\mathcal{I}(\mu|X)$$

whence Eq. (1.18) can be written as

$$\text{var}(T) \geq \cfrac{1}{nE\left\{\left(\frac{\partial}{\partial\mu}\log f_\mu(X)\right)^2\right\}}.$$

This leads to a natural definition of efficiency

$$\text{eff}(T) = \frac{1/\mathcal{I}(\mu|\pi^{(n)}oX)}{\text{var}(T)}$$

and we say that when an estimator T has efficiency one, then it is 100% efficient. For instance, assuming $f_\mu(x) = \phi(x - \mu)$, it is not hard to see $\mathcal{I}(\mu|X) = 1$ and that the sample mean or average carries efficiency of 100% which is the best possible. Consider the Tukey bisquare estimator, call it T_{BS}, having tuning constant $c = 4.685$. The asymptotic efficiency of this estimator at the standard normal distribution is 95%. Indeed the asymptotic efficiency of the sample median is $2/\pi \simeq 0.637$. The interpretation is that in large samples, using the sample mean requires only about 64% as many observations to estimate μ with a specified precision as would be required using the sample median. But to now compare two estimators, say the mean and the Tukey bisquare, when the underlying parametric distribution is $f_\mu(x) = (1 - \epsilon)\phi(x - \mu) + \frac{\epsilon}{3}\phi((x - \mu)/3)$ which is the density of a contaminated normal distribution, we can compare their relative efficiency, which is

$$\frac{\text{eff}T_{\text{BS}}}{\text{eff}\overline{X}} = \frac{\text{var}(\overline{X})}{\text{var}(T_{\text{BS}})}. \tag{1.19}$$

Clearly, this negates the need for evaluating the Fisher information. Also to be realistic we need, say, to decide how many iterations of the Newton–Raphson algorithm we have employed in evaluating the Tukey bisquare estimator starting

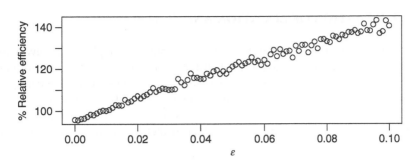

FIGURE 1.9   Estimated relative efficiency of the Tukey bisquare with tuning constant $c = 4.685$ relative to the sample mean based on 10 000 samples of size $n = 100$, with $v = 5$ Newton–Raphson iterations beginning from the median.

from the median, for example. We consider the number of Newton–Raphson iterations to be $v = 5$. By symmetry we can deduce that both estimators are unbiased for μ. Without loss of generality we can set $\mu = 0$. Since the estimators are unbiased and from any one sample we have an unbiased estimate $\hat{\mu}^2$ of the variance, sampling repeatedly 10 000 samples each of size $n = 100$, an estimate of the variance of the estimator $\hat{\mu}$ is gained from the SLLN, visualize $\hat{v}ar = \frac{1}{n}\sum_i \hat{\mu}_i^2$. This is done for epsilon fixed and for both of the estimates respectively from which an estimate of the relative efficiency at that epsilon is afforded. Then one calculates the estimates $\overline{X}$ and T_{BS} with values of epsilon incremented by 0.01 from zero to 0.1, and we obtain empirical measures of relative efficiency as indicated in Figure 1.9. Clearly, as the proportion of contamination increases from zero to 0.1, the efficiency of the Tukey bisquare estimator quickly overtakes that of the sample mean and indeed it appears to be roughly 140% when the proportion of contamination reaches 10%. Ten percent contamination is considered realistic and indeed Hampel et al. (1986) in summarizing their section 1.2c "The Frequency of Gross Errors" write "altogether, 1–10% gross errors in routine data seem to be more the rule rather than the exception."

Huber (1964) considered restricting distributions to be of the form $G(x) = (1 - \epsilon)\Phi(x) + \epsilon H(x)$, where H was only allowed to be a symmetric distribution. Tukey's results show that it is easy to find examples where estimators other than $\overline{X}$ can do better in terms of asymptotic performance at neighboring models to the normal distribution, while Huber includes a minimax solution that describes how to minimize the worst that nature can do to you by his choice of minimax distribution function, albeit a symmetric distribution function.

Much of the discussion in Chapter 2 is devoted to convergence results when G may be in an appropriate neighborhood of a parametric distribution. We explore ideas of convergence in distribution and further relate the generalities associated with convergence of the empirical curve K_{F_n} to K_G.

1.6   ESTIMATION OF LOCATION AND SCALE

Finally in this section we recognize that most practical applications involving M-estimators of location of a normal distribution use an estimate of scale of that distribution. Rarely in estimation do we assume a constant scale for the parametric distribution. The most practical and for that matter easily implementable scale estimate in the scenario where one needs to implement a robust M-estimator, such as the Tukey bisquare, is to choose the median absolute deviation about the median (MAD), defined as

$$\mathrm{MAD}(X_1, \ldots, X_n) = \mathrm{Med}\{|X_i - \mathrm{Med}(X1, \ldots, X_n)|\}.$$

This estimator uses the sample median twice, firstly, to get an estimate of location and then after forming the absolute residuals, finding the median of those absolute residuals. To make MAD consistent for σ, when assuming the data are $N(\mu, \sigma^2)$, we use

$$\mathrm{MADN} = \frac{\mathrm{MAD}}{0.6745}.$$

Then the Tukey bisquare estimate of location is found by replacing Eq. (1.7) by

$$\frac{1}{n} \sum_{i=1}^{n} \psi_{\mathrm{BS}} \left(\frac{X_i - \tau}{\mathrm{MADN}} \right) = 0.$$

Of course the M-estimation approach can also be taken in the estimation of location and scale. The estimating psi-function is a now a vector function $\boldsymbol{\Psi} = (\psi_1, \psi_2)^T$. Here ψ_1 governs the estimator for location and ψ_2 governs the estimator for scale. Both are implemented simultaneously through the vector equation

$$\boldsymbol{K}_n(\tau_1, \tau_2) = \frac{1}{n} \sum_{i=1}^{n} \boldsymbol{\Psi} \left(\frac{X_i - \tau_1}{\tau_2} \right) = \boldsymbol{0}. \tag{1.20}$$

A predecessor to the smooth (smooth meaning continuous first derivatives) location estimator of Beaton and Tukey (1974) was the Hampel three-part redescender first discussed in the book by Andrews et al. (1972). See also Hampel (1974). This estimator of location depends on three parameters a, b, c. For instance, the estimating function has the form

$$\psi_{1;a,b,c}(x) = \begin{cases} x & \text{for } 0 \leq |x| \leq a \\ a \, \mathrm{sign}(x) & \text{for } a \leq |x| \leq b \\ a\frac{c-|x|}{c-b}\mathrm{sign}(x) & \text{for } b \leq |x| \leq c \\ 0 & \text{for } c \leq |x| \end{cases} \tag{1.21}$$

See the illustration of Figure 1.10.

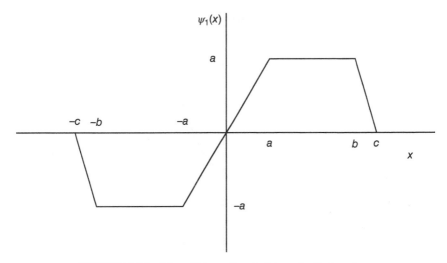

FIGURE 1.10   Plot of Hampel ψ_1 redescender for location.

However, it is also possible to construct an M-estimator of scale, along the lines of a three-part redescender in the following form.

$$\psi_{2;a,b,c}(x) = \begin{cases} x^2 - 1 - p & \text{for } 0 \leq |x| \leq a \\ a^2 - 1 - p & \text{for } a \leq |x| \leq b \\ (a^2 - 1 - p)\frac{c-|x|}{c-b} & \text{for } b \leq |x| \leq c \\ 0 & \text{for } c \leq |x| \, , \end{cases} \tag{1.22}$$

where p is defined implicitly from the equation $\int \psi_{2;a,b,c}(x)d\Phi(x) = 0$. See for example Clarke and Milne (2013). This function is illustrated in Figure 1.11. Then to obtain the estimator one solves Eq. (1.20) with $\Psi = \Psi_{a,b,c} = (\psi_{1;a,b,c}, \psi_{2;a,b,c})^T$.

It can now be noticed that the above functions have sharp corners, and consequently they do not have continuous derivatives. This means that expansions using the mean value theorem as in Eq. (1.11) are not now as straightforward, not to mention the fact that one now has a bivariate set of equations which requires multidimensional calculus. The extra dimension of scale also has to be incorporated. The calculus involved is more difficult, though it can be done using say the nonsmooth analysis discussed in Frank Clarke's book (1983b). This discussion is left till later chapters, but for a reference see Clarke (1986b).

Another important point is that Eq. (1.20) also admits multiple roots. Implementing such an equation requires a further understanding of consistency arguments which we begin in Chapter 2. Since both location and scale are estimated simultaneously, one might define a selection functional that combines both the location choice and a choice for scale. For instance, in estimating

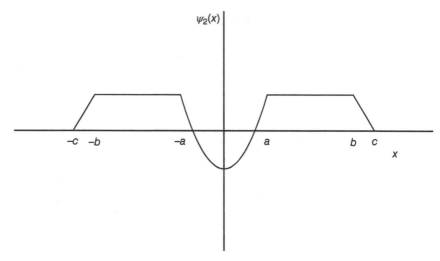

FIGURE 1.11   Plot of ψ_2 redescender for scale.

only scale of a normal distribution, essentially involving the second component of the vector equation (1.20) using the choice (1.22), one can choose to
select the scale estimator from possible multiple roots by using $\rho(F_n, \tau_2) =$
$|\{F_n^{-1}(3/4) - F_n^{-1}(1/4)\}/\{\Phi^{-1}(3/4) - \Phi^{-1}(1/4)\} - \tau_2|$. For location and scale
parameters estimated simultaneously, a useful selection functional to distinguish
the robust estimator from multiple roots of equations (1.20) is for example,

$$\rho(F_n; \tau_1, \tau_2) = (F_n^{-1}(1/2) - \tau_1)^2 +$$
$$(\{F_n^{-1}(3/4) - F_n^{-1}(1/4)\}/\{\Phi^{-1}(3/4) - \Phi^{-1}(1/4)\} - \tau_2)^2.$$
$$(1.23)$$

See, for example, section 7 of Clarke (1983a).

This choice of estimator, while in a sense rejects outliers in the tails of the
normal distribution, will have to be implemented carefully given possible multiple
roots of the equations. An alternative possibility, which has a longer history, is
to implement Huber's Proposal 2 estimator [see Huber (1964) and Huber (1981,
section 6.4)], where one down-weights but does not exclude outlying values. This
can be arrived at by setting $b = c = \infty$ and for purposes of discussion $k = a$. Then
the estimator for location and scale looks like $\Psi = (\psi_1, \psi_2)^T$ where

$$\left.\begin{array}{l} \psi_1(x) = \min(|x|, k)\text{sign}(x) \\ \text{and} \\ \psi_2(x) = \psi_1^2(x) - E_\Phi[\psi_1^2(Z)] \end{array}\right\} \quad -\infty < x < \infty \qquad (1.24)$$

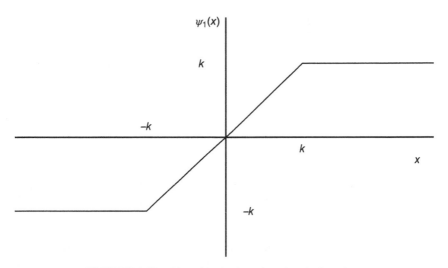

FIGURE 1.12   Plot of Huber's ψ-function for location.

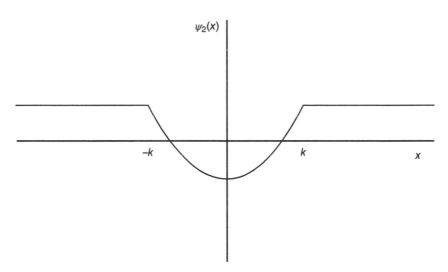

FIGURE 1.13   Plot of Huber's χ-function for estimating scale.

and the graphs of ψ_1 and ψ_2 are in Figures 1.12 and 1.13 respectively. The implementation of Huber's Proposal 2 is discussed in Venables and Ripley (2002) for example. In a novel approach to gaining a redescending estimator for location and scale, where the Ψ-function is continuously differentiable for both component functions, Bachmaier (2007) in essence provides the following functions:

$$\psi_1(x) = \begin{cases} -\psi_1(-x) & \text{for} \quad x < 0 \\ x & \text{for } 0.0 \le x \le 0.9 \\ -(x - 1.4)^2 + 1.15 & \text{for } 0.9 < x < 1.9 \\ -x + 2.8 & \text{for } 1.9 \le x \le 2.3 \\ 0.5(x - 3.3)^2 & \text{for } 2.3 < x < 3.3 \\ 0 & \text{for } 3.3 \le x \end{cases} \tag{1.25}$$

$$\psi_2(x) = \begin{cases} x^2 - 0.7 & \text{for} \quad |x| \le 1.0 \\ -(|x| - 2)^2 + 1.3 & \text{for } 1.0 \le |x| \le 3.0 \\ (3.3 - (|x|)^2/0.3 & \text{for } 3.0 \le |x| \le 3.3 \\ 0 & \text{for } 3.3 \le |x| \quad . \end{cases} \tag{1.26}$$

Bachmaier (2007) discusses in detail various ways of arriving at a solution that is consistent. With either choice of redescender involving Hampel's choice and the adapted Hampel redescender for scale, or Bachmaier's choice, when one solves Eq. (1.20) one needs to search for solutions using iterative nonlinear equation solving algorithms, which require initial estimates of location and scale (Bachmaier's functions are illustrated in Figures 1.14 and 1.15). One can use Huber's estimates or the estimates based on nonparametric quantities such as $(\text{Median}, \text{MADN})^T$ for the initial estimates. In small samples, it may be worth searching say starting with a grid of initial estimates and then implementing the selection functional to choose the final estimate from possible multiple solutions. While this entails extra

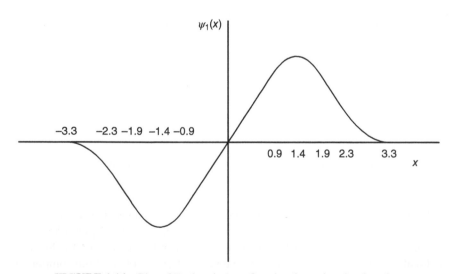

FIGURE 1.14   Plot of Bachmaier's ψ-function for estimating location.

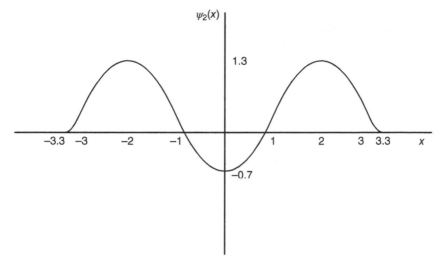

FIGURE 1.15   Plot of Bachmaier's ψ_2-function for estimating scale.

work, it may be philosophically more satisfying in that one gives zero weight to observations that are in the tails of the distribution. Extreme outliers are thus given zero weight in the estimation.

The problems thrown up by the introduction of redescending M-estimates of location and scale are found more widely in parametric estimation and even in the restricted scenario of maximum likelihood estimation. The following are the questions often raised: How do we numerically find a solution? If there are more than one solution to the problem (this may involve more than one solution to the maximizing equations, or indeed in some parametric models different solutions that give equal maxima to the likelihood), how do we distinguish between solutions? In Chapter 2 we give a general theory of consistency, asymptotic normality, and asymptotic efficiency based on a functional approach to estimation.

PROBLEMS

1.1. A collection $\tilde{\mathcal{A}}$ of subsets of Ω is called a σ-field if it satisfies the following conditions:

(i) $\emptyset \in \tilde{\mathcal{A}}$

(ii) if $A_1, A_2, \ldots \in \tilde{\mathcal{A}}$ then $\cup_{i=1}^{\infty} A_i \in \tilde{\mathcal{A}}$

(iii) if $A \in \tilde{\mathcal{A}}$ then $A^c \in \tilde{\mathcal{A}}$, where A^c is the complement of A.

Use elementary set operations to show that if $\tilde{\mathcal{A}}$ is closed under countable intersections: that is, if $A_1, A_2, \ldots$ are in $\tilde{\mathcal{A}}$, then so is $\cap_i A_i$.

1.2. Consider a die that results in six with probability one in six. Consider a sequence of independent tosses of the die where for the ith the random variable X_i is a one if the die is a six and a zero otherwise. The random variables have a common distribution. Show that the mean of this distribution is $\frac{1}{6}$ and conclude what is the limiting value in probability and also almost surely of the sample average $\frac{1}{n}\sum_{i=1}^{n}X_i$ as $n \to \infty$.

1.3. Consider a random variable with the exponential distribution which is defined on the positive real line and has density $f_\theta(x) = \exp(-x/\theta)/\theta$. Find the expectation and variance of this random variable.

1.4. Show that the class of functions $\mathcal{A} = \{\psi_{BS}(\cdot - \tau) : \tau \in \mathcal{E}\}$ is equicontinuous.

1.5. Consider a sequence of independent coin tosses of a fair coin and let $\overline{X}_n$ the average number of heads. Show that the distribution of the random variable $\sqrt{n}\left(\overline{X}_n - \frac{1}{2}\right)$ converges to a $N\left(0, \frac{1}{4}\right)$ random variable.

1.6. Let $X_1, X_2, \ldots$ be an *i.i.d.* sequence of observations with common distribution F that has a symmetric density f which is located about $\mu = F^{-1}\left(\frac{1}{2}\right)$. Under various regularity conditions the asymptotic variance of the median is

$$\text{var}[\text{Med}(X_1, \ldots, X_n)] \sim \frac{1}{4nf^2\left(F^{-1}\left(\frac{1}{2}\right)\right)}.$$

Show that when f is the normal density with unit scale that the efficiency of the median is $\frac{2}{\pi}$ and this does not depend on μ. Also investigate whether or not the efficiency depends on the scale of the normal density σ say?

1.7. Consider the following data: 1.0, −1.8, −0.5, −1.4, 0.7, 0.9, −0.6, 1.9, −0.1, 3.5. Calculate the median and MADN for these data. Subsequently, write an algorithm using the Newton–Raphson iteration of formula (1.8) by substituting the Tukey bisquare function (1.6) to evaluate the Tukey bisquare estimator when using MADN to estimate the scale.

1.8. For the data of the previous question and using Bachmaier's choice of M-estimator as defined by Eqs. (1.25) and (1.26) use a nonlinear equation solver starting the iteration from (Median, MADN). Investigate whether there are other roots of the equation by forming a grid of initial estimates by choosing the location estimate to vary from −1 to 1 in steps of 0.2 and scale to vary from 0.5 to 2 in steps of 0.5 and iterating to a root. This will involve running the algorithm several times. If there are multiple roots, decide which is the closest to (Median, MADN)?

2

THE FUNCTIONAL APPROACH

2.1 ESTIMATION AND CONDITIONS A

A definition of estimators as functions of the empirical distribution function and the relationship of the functional approach to estimation are given here. von Mises (1947) is recognized as one of the first proponents of describing estimators as functionals. For instance a solution of Eq. (1.7), provided it is well defined, can be thought of as a functional mapping from $\mathcal{G}$ to the parameter space $\Theta \subset \mathcal{E}^r$. Knowing properties of the general map $T : \mathcal{G} \to \Theta$ can reveal much that is to be known about an estimator $T_n(X_1, \ldots, X_n) = T[F_n]$ at least for independent identically distributed sequences of random variables. This functional approach will be observed throughout the book from here. An explanation with some history of likelihood-based approaches is given with notions of efficiency in the parametric setting and how and why they can sometimes succeed and more often do badly fail under "small" deviations from the parametric model. A discussion of "small" deviations is given in light of the neighborhoods that could theoretically truly reflect the underlying distribution, even though we have a model parametric distribution in mind. As we have seen in Chapter 1, continuity and differentiability of statistical functionals, in a sense yet to be defined, can lead to consistency and asymptotic normality of resultant estimators defined from the aforesaid functionals.

Robustness Theory and Application, First Edition. Brenton R. Clarke.
© 2018 John Wiley & Sons, Inc. Published 2018 by John Wiley & Sons, Inc.
Companion website: www.wiley.com/go/clarke/robustnesstheoryandapplication

Continuity of the statistical functional also has an important part to play, not least because of consistency of an estimator. However, if our statistical functional is continuous with respect to a metric distance called the Prohorov distance, all the attendant implications of robustness follow. See Hampel (1971) for example. To relate these notions we need firstly to describe in detail what a metric distance between say two distribution functions in the space of distributions $\mathcal{G}$ is. A metric on the set $\mathcal{G}$ is a distance $d : \mathcal{G} \times \mathcal{G} \to \mathcal{E}$. For all $F, G, H \in \mathcal{G}$ this distance is required to satisfy the following conditions:

$d(F, G) \geq 0$;

$d(F, G) = 0$ if and only if $F = G$;

$d(F, G) = d(G, F)$;

$d(F, G) \leq d(F, H) + d(H, G)$.

Some very useful distances are the Kolmogorov distance (also referred to as the Kolmogorov–Smirnov distance) defined for distributions on the real line $d_k(F, G) = \sup_{-\infty < x < +\infty} |F(x) - G(x)|$, and the Prohorov distance defined on distributions on the more general space $(R, \mathcal{B})$, where $\mathcal{B}$ are the Borel sets of the metric space R with metric $\tilde{d}$ say. Here the Prohorov distance is defined as

$$d_{\mathrm{p}}(F, G) = \inf\{\delta | F[A] \leq G[A^\delta] + \delta, G[A] \leq F[A^\delta] + \delta, \quad \text{for all} \quad A \in \mathcal{B}\}.$$

Here A^δ is the closed δ-neighborhood of set A. The Lévy distance, which we shall denote by d_{L}, is defined for distributions on the real line by restricting the sets A in the above formula to be of the form $(-\infty, x]$, $x \in \mathcal{E}$.

One can imagine that while one believes that there is a true distribution $F_\theta \in \{F_\tau(x) | \tau \in \Theta \subset \mathcal{E}^r\}$ which is hopefully going to describe the physical reality of an experimental problem, it is quite clear that in most situations the data are not exactly normal, or exponential for instance, but in fact they may differ in their true distribution because of the following realities, well quoted in the robustness literature:

1. Rounding and grouping of data: even with continuous data we only ever consider a finite number of decimal points.
2. The occurrence of "gross errors" (such as blunders in the measuring, missed decimal points, miss-punching of observations into the computer).
3. The model may have been conceived only as an approximation in the first place. The assumption of Gaussianity is usually wrong and tails tend to be fatter (heavier) than normal tails.
4. The data may not be independent. There exist many situations where data can be autocorrelated in nature, but that autocorrelation is not recognized.

Clearly, small changes in (1) are reflected in small changes in the Prohorov or Lévy distance metrics, while (2) and (3) can be reflected in small changes in the

Kolmogorov, Lévy, and Prohorov distance metrics. In a sense then we seek estimating functionals $T[\cdot] : \mathcal{G} \to \Theta$ that are at least continuous with respect to such metrics. Suffice to say, many classical estimates of the maximum likelihood type based on parametric families with exponentially decreasing tails are not continuous. Tentative results on (4) are relayed in Bednarski (2010). See Hampel et al. (1986) section 1.2b for a more detailed discussion of deviations from parametric models with history.

To give here a mathematical discussion with more generality, we consider maximum likelihood type equations of the form

$$K_{F_n}(\tau) = \frac{1}{n} \sum_{i=1}^{n} \Psi(X_i, \tau) = 0, \qquad (2.1)$$

where $X_1, \ldots, X_n$ are independent identically distributed random variables taking values in a separable metrizeable space R, and Ψ is an $r \times 1$ vector function with domain $R \times \Theta$ which has a continuous partial derivative $(\partial/\partial\tau)\Psi(x, \tau)$. Estimators that are solutions of Eq. (2.1) are generally termed M-estimators and include the maximum likelihood estimator (MLE) and some minimum distance estimators. Somewhat surprisingly, given the volume of works related to these estimators, it can also include the method of moments estimator, which has a longer history even than MLEs. But with the emergence of robust estimators it is often better to choose something that is not optimal, but nevertheless performs well at the parametric model, and can do well also in neighborhoods of the parametric model. The Tukey bisquare estimating M-functional is such an estimator, though notably not the only one.

A single functional root of Eq. (2.1) may be written $T[\Psi, F_n]$ where F_n is the empirical distribution function. More generally $T[\Psi, G]$ can be defined as a functional root of equations

$$K_G(\tau) = \int_R \Psi(x, \tau) dG(x) = 0. \qquad (2.2)$$

$T[\Psi, G] = +\infty$ if no root exists. In the event of several roots of Eq. (2.2), a distance criterion ρ_o is employed to select the estimator from them. Classically, the roots correspond to extrema of the distance, but this is not assumed here. Specifically, the general functional $T[\Psi, \rho_o, \cdot]$ is defined as the solution to

$$\inf_{\tau \in S(\Psi, G)} \rho_o(G, \tau) = \rho_o(G, T[\Psi, \rho_o, G]), \qquad (2.3)$$

where

$$S(\Psi, G) = \left\{ \tau \in \Theta \mid \int_R \Psi(x, \tau) dG(x) = 0 \right\}, \qquad (2.4)$$

if a solution exists. Otherwise $T[\Psi, \rho_o, G] = +\infty$. Conditions on both Ψ and ρ_o determine consistency, the very minimum requirement being Fisher consistency, which means that $T[\Psi, \rho_o, F_\theta] = \theta$ for all $\theta \in \Theta$. That is, if you have the whole population you must get the corresponding true parameter. This implicitly also assumes the parametric model for the population is identifiable. A parametric family $\mathcal{F}$ is identifiable if and only if $F_{\theta_1} = F_{\theta_2}$ implies $\theta_1 = \theta_2$ for all $\theta_1, \theta_2 \in \Theta$. That is, two different parameters cannot give the same distribution.

One of the most studied estimators is the MLE. Assuming that there is a probability density function $f_\tau(x)$ which corresponds to the cumulative parametric distribution $F_\tau \in \mathcal{F}$, the MLE, at least for observations defined on a subset of $\mathcal{E}^k$, is governed by the choice of the efficient score function

$$\Psi(x, \tau) = \frac{\frac{\partial}{\partial \tau} f_\tau(x)}{f_\tau(x)} \tag{2.5}$$

and

$$\rho_o(G, \tau) = - \int_R \log f_\tau(x) dG(x) \tag{2.6}$$

if a solution exists. Otherwise $T[\Psi, \rho_o, G] = \infty$. There are some implicit assumptions about the regularity of the density function, for instance that its support does not have a boundary which involves τ, and further assumptions on Ψ and ρ_o that need to be checked before one can sensibly use them.

Even Kallianpur and Rao (1955) in interpreting notes of Fisher mistakenly asserted, and later retracted, that only analytic functions of the frequencies were considered, which is a harsh condition to impose, in order to obtain "efficiency." It is the form of Ψ that governs questions of efficiency, which we shall see below.

The key in Chapter 1 to the convergence of $T[\psi_{BS}, F_n]$ to $T[\psi_{BS}, F_\theta]$ was to note, for example, that $K_n(\tau) = \int_{-\infty}^{+\infty} \psi_{BS}(x - \tau) dF_n(x)$ converged to $K_{F_\theta}(\tau)$ uniformly in τ. This also can be said for the derivatives of these functions, that is, $K'_n(\tau)$ converged to $K'_{F_\theta}(\tau)$. As the empirical distribution function becomes closer to the parametric model distribution function, then so does the empirical curve get closer to the asymptotic curve. This continuity is generalized in a result that is relayed in terms of the Prohorov metric distance in the following, potentially abstract result, but one that has wide implications.

Theorem 2.1: *(Clarke, 1983a, Theorem 6.1) Let $\mathcal{A}$ be a class of continuous functions on separable metrizeable space R possessing the following two properties: (1) $\mathcal{A}$ is uniformly bounded, that is, there exists a constant H such that $|f(x)| \leq H < +\infty$ for all $f \in \mathcal{A}$ and $x \in R$; and (2) $\mathcal{A}$ is equicontinuous. Let $F_\theta \in \mathcal{G}$ be given. Then* *for every* $\delta > 0$ *there is an* $\epsilon > 0$ *such that*

$$d_\mathrm{p}(G, F_\theta) \le \epsilon \quad implies\ \sup_{f \in \mathcal{A}} \left| \int_R f \mathrm{d}G - \int_R f \mathrm{d}F_\theta \right| < \delta. \tag{2.7}$$

Proof. See Appendix A. In remarks following the proof of this theorem (see Clarke 1983), it is noted that if $R = \mathcal{E}$ the result also follows through using d_k and d_L.

The class $\mathcal{A}$ is in a sense the most general. A weaker condition than (1) is to assume $\sup_{f \in \mathcal{A}} \int_R |f| \mathrm{d}F_\theta = m < +\infty$ but allows functions in $\mathcal{A}$ to be unbounded. By choosing $\{f_n\} \subset \mathcal{A}$, $\{y_n\}$ so that $|f_n(y_n)| \to +\infty$ as $n \to +\infty$, consider for any $\epsilon > 0$, $G_n = (1 - \epsilon)F_\theta + \epsilon\delta_{y_n}$

$$\sup_{f \in \mathcal{A}} \left| \int f \mathrm{d}G_n \int f \mathrm{d}F_\theta \right| > \epsilon|f_n(y_n) \int_R f_n \mathrm{d}F_\theta| \tag{2.8}$$

$$> \epsilon(|f_n(y_n)| - m) \to +\infty \text{ as } n \to \infty. \tag{2.9}$$

This violates Eq. (2.7) since $d_\mathrm{p}(F_\theta, G_n) \le \epsilon$.

If (2) does not hold, there is a $\delta > 0$ and $x \in R$ and a sequence $\{y_n\}, y_n \to x$ as $n \to \infty$, so that $\sup_{f \in \mathcal{A}} |f(x) - f(y_n)| > \delta$. Suppose at θ, $F_\theta = \delta_x$, then $d_\mathrm{p}(F_0, G_n) \to 0$ as $n \to \infty$ but $\sup_{f \in \mathcal{A}} |\int_R f \mathrm{d}F_0 - \int_R f \mathrm{d}G_n| > \delta$, contradicting Eq. (2.7).

By considering neighborhoods of a distribution $G \in \mathcal{G}$ for which the ordering property $0 < \epsilon_1 \le \epsilon_2$ implies $n(\epsilon_1, G) \subset n(\epsilon_2, G)$ holds, we can now lay down conditions by which one may achieve results of consistency and asymptotic normality of estimators $T[\Psi, \rho_o, \cdot]$. Of course neighborhoods can easily be generated by a metric distance by writing $n(\epsilon, G) = \{F \in \mathcal{G} | d(F, G) < \epsilon\}$ or they may be of the form of an ϵ-contaminated neighborhood $n(\epsilon, G) = \{(1 - \epsilon)G + \epsilon F | F \in \mathcal{G}\}$. The following conditions are from Clarke (1983a):

Conditions A

A_0: $T[\Psi, \rho, F_\theta] = \theta$.

A_1: Ψ is an $r \times 1$ vector function on $R \times \Theta$ and has continuous partial derivatives on $R \times D$, where $D \subset \Theta$ is some nondegenerate compact interval containing θ in its interior.

A_2: $\{\Psi(x, \tau)|\tau \in D\}$, $\{\frac{\partial}{\partial \tau}\Psi(x, \tau)|\tau \in D\}$ are bounded above in Euclidean norm ($|A| = \{\mathrm{trace}(A^T A)\}^{1/2}$) by some continuous function g that is integrable with respect to all $G \in n(\epsilon, F_\theta)$ for some $\epsilon > 0$.

A_3: The matrix

$$M(\theta) = \int_R \{\frac{\partial}{\partial \tau}\Psi(x, \theta)\}dF_\theta(x)$$

is nonsingular.

A_4: Given $\delta > 0$ there exists an $\epsilon > 0$ such that for all $G \in n(\epsilon, F_\theta)$
$\sup_{\tau \in D}|\int_R \Psi(x, \tau)dG(x) - \int_R \Psi(x, \tau)dF_\theta(x)| < \delta$, and
$\sup_{\tau \in D}|\int_R \frac{\partial}{\partial \tau}\Psi(x, \tau)dG(x) - \int_R \frac{\partial}{\partial \tau}\Psi(x, \tau)dF_\theta(x)| < \delta$.

Conditions **A** are sufficiently general to either establish a weakly continuous root to Eq. (2.2), continuous with respect to the Prohorov distance metric, by showing existence of an asymptotically unique consistent solution in shrinking small neighborhoods of θ, or by relaxing the neighborhoods of F_θ one can show existence of an asymptotically unique consistent root that converges almost surely to θ, assuming F_θ is the distribution that generates the sequence Eq. (1.1). Since a first step is to produce existence of an asymptotically unique root in a neighborhood of θ, we use an auxiliary selection functional $\rho(G, \tau) = |\tau - \theta|$. The subsequent $T[\Psi, \rho, G]$ using the auxiliary functional is then used to discover the properties of $T[\Psi, \rho_o, G]$ for some suitably chosen ρ_o, although of course its properties are immediately applicable if only one solution to the equations exists for then $T[\Psi, \rho, \cdot]=T[\Psi, \cdot]$.

We first begin with a lemma (Clarke 1983a, Lemma 3.2). This lemma shows that under assumptions of conditions **A** there exists a ball of the parameter space of size $\delta_1 > 0$ about θ and a neighborhood of distributions about F_θ where the gradient matrix is non-singular. This seemingly innocuous result is made use of in the proof of consistency and weak continuity of estimating functionals that are roots of Eq. (2.1) or (2.2). We repeat the arguments of Clarke (1983a).

Lemma 2.1: *Let conditions A hold for some* Ψ, ρ. *Then there is a $\delta_1 > 0$ and an $\epsilon_1 > 0$ such that for every $\tau \in \cup_{\delta_1}(\theta)$ the open ball of radius δ_1 and center θ, and every $G \in n(\epsilon_1, F_\theta)$ the matrix*

$$M(\tau, G) = \int_R \{\frac{\partial}{\partial \tau}\Psi(x, \tau)\}dG(x)$$

is non-singular.

Proof: By continuity of a determinant as a function of the elements of a matrix, choose $\eta > 0$ such that $||A - M(\theta)|| < \eta$ implies $|\det\{A\}| > \frac{1}{2}|\det\{M(\theta)\}|$ for an $r \times r$ matrix A. Assumptions A_1, A_2 imply $M(\tau, F_\theta)$ is continuous in $\tau \in D$. So choose $\delta_1 > 0$ such that $\tau \in \cup_{\delta_1}(\theta) \subset D$ implies $||M(\tau, G) - M(\tau, F_\theta)|| < \eta/2$. The lemma is proved by the triangle inequality of norms.   □

So just as in Chapter 1 where it is seen that the derivative of the empirical curve formed from using the Tukey bisquare remains nonzero in a region about the true parameter which leads to arguments of asymptotically unique consistent roots, now multivariate arguments using the lemma immediately above lead to a theorem which is much more general but has similar implications.

The following theorem is from analysis and is used in the proof of continuity and/or consistency of the functional $T[\Psi, \rho, \cdot]$ in Theorem 2.3. The proof of Theorem 2.3 mimics a proof of an asymptotically unique consistent root of the maximum likelihood equations which is given by Foutz (1977), but with the aid of conditions **A** given earlier we now have ways where the conditions of Foutz (1977) can be realized. The proofs of consistency extend to the class of M-estimators where with appropriate choice of neighborhoods the functions Ψ can be unbounded in the observation space variable. However, more importantly the general approach can be used to show weak continuity of M-estimators for functions Ψ that are known to be bounded with continuous and bounded partial derivatives. This has important robustness ramifications and helps us delineate between estimates that are just consistent at the parametric model and those that are robust and consistent even in neighborhoods of the parametric model.

Theorem 2.2: *(Inverse Function Theorem) Suppose f is a mapping from Θ into $\mathcal{E}^r$, the partial derivatives exist and are continuous on Θ, and the matrix of derivatives of f exist and are continuous on Θ, and the matrix of derivatives $f'(\theta^*)$ has inverse $f'(\theta^*)^{-1}$ for some $\theta^* \in \Theta$. Write $\lambda = 1/(4\|f'(\theta^*)^{-1}\|)$. Use the continuity of the elements of $f'(\tau)$ to fix a neighborhood $\cup_\delta(\theta^*)$ of sufficiently small radius $\delta > 0$ to ensure that $\|f'(\tau) - f'(\theta^*)\| < 2\lambda$ whenever $\tau \in \cup_\delta(\theta^*)$. Then*

(a) *for every $\tau_1, \tau_2 \in \cup_\delta(\theta^*)$ $\|f(\tau_1) - f(\tau_2)\| \geq 2\lambda \|\tau_1 - \tau_2\|$; and*

(b) *the image set $f(\cup_\delta(\theta^*))$ contains the open neighborhood with radius $\lambda\delta$ about $f(\theta^*)$.*

Conclusion (a) ensures that f is one-to-one on $\cup_\delta(\theta^*)$ and that f^{-1} is well defined on the image set $f(\cup_\delta(\theta^*))$.

Remark 2.1: The $\|A\|$ can also be interpreted as the least upper bound of all numbers $\|Ay\|$, where y ranges over all the vectors in $\mathcal{E}^r$ with $\|y\| \leq 1$; cf. Foutz (1977).

Theorem 2.3: *(Clarke, 1983a, Theorem 3.2) Let $\rho(G, \tau) = |\tau - \theta|$ and suppose conditions A hold. Then given $\kappa > 0$ there exists an $\epsilon > 0$ such that $G \in n(\epsilon, F_\theta)$ implies $T[\Psi, \rho, G]$ exists and is an element of $\cup_\kappa(\theta)$. Further for this ϵ there is a $\kappa^* > 0$ such that*

$$S(\Psi, G) \cap \cup_{\kappa^*}(\theta) = T[\Psi, \rho, G], \qquad (2.10)$$

and $M(\tau, G)$ is non-singular for $\tau \in \cup_{\kappa^*}(\theta)$. For any null sequence of positive numbers $\{\epsilon_k\}$, let $\{G_k\}$ be any arbitrary sequence for which $G_k \in n(\epsilon_k, F_\theta)$. Then

$$\lim_{k \to \infty} T[\Psi, \rho, G_k] = T[\Psi, \rho, F_\theta] = \theta. \qquad (2.11)$$

Remark 2.2: Theorem 2.3 demonstrates the uniqueness of a solution of Eq. (2.2) in a region $\cup_{\kappa^*}(\theta)$ by Eq. (2.10) and continuity of the functional by Eq. (2.11).

Proof. (of Theorem 2.3) Write $\lambda = 1/(4||M(\theta)^{-1}||)$. By continuity of $M(\tau, F_\theta)$ in τ, choose $0 < \kappa^* < \min(\delta_1, \kappa)$ such that $\tau \in \cup_{\kappa^*}(\theta)$ implies $||M(\tau, F_\theta) - M(\theta)|| < \lambda/2$. Here δ_1, ϵ_1 are given by Lemma 2.1. For $G \in n(\epsilon_1, F_\theta)$ define $\lambda(G) = 1/(4||M(\theta, G)^{-1}||)$. Choose $0 < \epsilon^* \le \epsilon_1$ so that

$$||M(\tau, G) - M(\theta, G)|| \le ||M(\tau, G) - M(\tau, F_\theta)||$$

$$+ ||M(\theta, G) - M(\theta)|| + ||M(\tau, F_\theta) - M(\theta)||$$

$$\le \lambda < 2\lambda(G) \quad \text{whenever} \quad G \in n(\epsilon^*, F_\theta)$$

for all $\tau \in \cup_{\kappa^*}(\theta)$. Note A_1, A_2 imply $K_G(\tau)$ has continuous partial derivatives $M(\tau, G)$. Properties (a) and (b) ensure $K_G(\cdot)$ is a one-to-one function from $\cup_{\kappa^*}(\theta)$ onto $K_G(\cup_{\kappa^*}(\theta))$ and that the image set contains the open ball of radius $\lambda\kappa^*/2$ about $K_G(\theta)$. Now choose $0 < \epsilon' \le \epsilon^*$ such that

$$|K_G(\theta) - 0| < \lambda\kappa^*/2.$$

Then it is clear that $0 \in K_G(\cup_{\kappa^*}(\theta))$ for all $G \in n(\epsilon^*, F_\theta)$ and that the image set contains the open ball of radius $\lambda\kappa^*/2$ about $K_G(\theta)$. Consider the inverse function

$$K_G^{-1} : K_G(\cup_{\kappa^*}(\theta)) \to \cup_{\kappa^*}(\theta) \text{ for } G \in n(\epsilon', F_\theta).$$

It is well defined whenever $K_G(\tau)$ is one-to-one. Since $0 \in K_G(\cup_{\kappa^*}(\theta))$ for $G \in n(\epsilon', F_\theta)$, we conclude that with $\epsilon' = \epsilon$ there exists a unique root of equations (2.2) in $\cup_{\kappa^*}(\theta)$ whenever $G \in n(\epsilon, F_\theta)$. That is, Eq. (2.10) holds. If we let $\{\kappa_i^*\}_{i=1}^\infty$ be a null sequence for which $0 < \kappa_i^* \le \kappa^*$, $i \ge 1$, there exists a corresponding sequence of $\{\epsilon'\}$. Since $\{\epsilon_n\}$ is null, there is some $j(i)$ for which $\epsilon_{j(i)} \le \epsilon_i'$, whenever $G_{j(i)} \in n(\epsilon_i', F_\theta)$, $i \ge 1$. Hence,

$$T[\Psi, \rho, G_k] = K_{G_k}^{-1}(0) \cap \cup_{\kappa^*}(\theta) \quad k > \inf_{1 \le i < \infty} j(i),$$

and

$$\lim_{k \to \infty} T[\Psi, \rho, G_k] = T[\Psi, \rho, F_\theta] = \theta.$$

$\square$

These results are local results for local solutions of the estimating equations (2.1). But many statistical problems need a global solution, which is afforded by use of a selection functional ρ_o. This separation of local consistency and global consistency is given by the following theorem.

Theorem 2.4: *(Clarke, 1983a, Theorem 4.1) Assume conditions A hold for the functional $\rho(G, \tau) = |\tau - \theta|$. Suppose $\rho_o(G, \tau)$ is a selection functional such that every neighborhood N of θ*

$$\inf_{\tau \notin N} \rho_o(F_\theta, \tau) - \rho_o(F_\theta, \theta) > 0, \tag{2.12}$$

and for every $\eta > 0$ there exists an ϵ such that $G \in n(\epsilon, F_\theta)$ implies $\rho_o(G, \tau)$ is continuous in $\tau \in \Theta$ and satisfies

$$\sup_{\tau \in \Theta} |\rho_o(G, \tau) - \rho_o(F_\theta, \tau)| < \eta. \tag{2.13}$$

Then for any given $\kappa > 0$ there exists an ϵ_0 such that $G \in n(\epsilon_0, F_\theta)$ implies $T[\Psi, \rho_o, G]$ exists, is unique, and lies in $\cup_\kappa(\theta)$.

Proof. From Theorem 2.3 there is $0 < \kappa^* < \kappa$ and $\epsilon > 0$ such that $G \in n(\epsilon, F_\theta)$ implies Eq. (2.10) holds. Denote

$$\delta(\kappa^*) = \inf\{\rho_o(F_\theta, \tau) - \rho_o(F_\theta, \theta) | \tau \in \Theta - \cup_{\kappa^*}(\theta)\}. \tag{2.14}$$

Choose $0 < \kappa' < \kappa^*$ so that $\tau \in \cup_{\kappa'}(\theta)$ implies

$$|\rho_o(F_\theta, \tau) - \rho_o(F_\theta, \theta)| < \frac{\delta(\kappa^*)}{2}.$$

For $\kappa' > 0$ choose $0 < \epsilon_o \leq \epsilon$ so that $G \in n(\epsilon_o, F_\theta)$ implies $T[\Psi, \rho, G] \in \cup_{\kappa'}(\theta)$ and

$$\sup_{\tau \in \Theta} |\rho_o(G, \tau) - \rho_o(F_\theta, \tau)| < \frac{\delta(\kappa^*)}{4}.$$

Note that Eq. (2.10) remains true for $G \in n(\epsilon_o, F_\theta)$. Then

$$\rho_o(G, T[\Psi, \rho, G]) < \rho_o(F_\theta, T[\Psi, \rho, G]) + \frac{\delta(\kappa^*)}{4}$$

$$< \rho_o(F_\theta, \theta) + \frac{3\delta(\kappa^*)}{4}$$

$$< \rho_o(F_\theta, \tau) - \frac{\delta(\kappa^*)}{4} \quad \text{uniformly in } \tau \in \Theta - \cup_{\kappa^*}(\theta)$$

$$< \rho_o(G, \tau) \quad \text{uniformly in } \tau \in \Theta - \cup_{\kappa^*}(\theta).$$

Hence,

$$\inf_{\tau \in S(\Psi, G)} \rho_o(G, \tau) = \rho_o(G, T[\Psi, \rho, G]),$$

and

$$T[\Psi, \rho_o, G] = T[\Psi, \rho, G] \in \cup_{\kappa'}(\theta) \subset \cup_{\kappa}(\theta).$$

$\square$

It can be remarked that Eq. (2.12) is often satisfied when ρ_o is chosen to be a distance, for example,

$$\rho_o(F_n, \tau) = \int (F_n(x) - F_\tau(x))^2 dW(x) \qquad (2.15)$$

for suitable weight functions W. Minimizing Eq. (2.15) can be shown to give Eq. (2.1) [see Boos (1981) and Clarke and Heathcote (1994) for example]. Here assumption (2.12) is equivalent to

$$\inf_{\tau \notin N} \int (F_\theta(x) - F_\tau(x))^2 dW(x) > 0 \qquad (2.16)$$

which for usual choices of W, including Lebesgue measure and exponential weight functions, can be shown to be a result of identifiability of the parametric family. Assumption (2.16) is found to be in the minimum distance theory of Pollard (1980) and also Wolfowitz (1957). On the other hand, the MLE is adopted into the framework of the selection functional with the choice of both (2.5) and (2.6). Assumption (2.12) is

$$\inf_{\tau \notin N} E_{F_\theta}[-\log f_\tau(X)] - E_{F_\theta}[-\log f_\theta(X)] > 0$$

or equivalently

$$\sup_{\tau \notin N} E_{F_\theta}[\log f_\tau(X)] < E_{F_\theta}[\log f_\theta(X)].$$

In comparison to the global consistency argument of Wald (1949), this assumption is similar to Lemma 1 of that paper which shows under suitable conditions on $\{f_\tau\}$ for $\tau \neq \theta$ that

$$E_{F_\theta}[\log f_\tau(X)] < E_{F_\theta}[\log f_\theta(X)].$$

When $\rho(F_\theta, \tau)$ is continuous in τ the two statements are equivalent.

Finally, it can be remarked that while assumption (2.13) is a sufficient condition for the continuity result, it may not be necessary. If we consider a Fréchet space R and set $f(G)$ to be $\int_R ||x|| dG(x)$ when this integral is finite. Otherwise put $f(G) = 0$. Clearly, using the selection functional $\rho_o(G, \tau)$ of Theorem 2.4 is equivalent to using $\rho_1(G, \tau) = \rho_o(G, \tau) + f(G)$, but the latter need not satisfy Eq. (2.13).

2.2 CONSISTENCY

Classically, arguments for consistency have had two threads, one is local consistency, the other global consistency, both assuming that one has Fisher consistency. Several authors including Cramér (1946), Huzurbazaar (1948), Tarone and Gruenage (1975), and Foutz (1977) examine consistency arguments for local solutions θ_n^* of maximum likelihood equations. They consider both existence and uniqueness of local solutions of the equations (2.1) when Ψ is the efficient score function. There may be other solutions $\{\tilde{\theta}_n\}$ which are not consistent. Global arguments for consistency using Eq. (2.6) where one finds that parameter that maximizes the likelihood are given in Wald (1949), while for M-estimators consider Huber (1967). Similar arguments are found in minimum distance estimation, where it is the extremum value that is the estimator, for instance see Wolfowitz (1957). Most of these arguments assume that the estimator $T[F_n]$ is being studied when F_n is generated by F_θ. Under this assumption we are able to elucidate verifiable conditions by which one can obtain existence of an aysmptotically unique consistent root.

Consider now the following distribution neighborhoods: for $\delta > 0$, define for a nondegenerate compact set $D \subset \Theta$ containing θ in its interior, let

$$n(\delta, F_\theta) = \{G \in \mathcal{G}|$$

$$\sup_{\tau \in D} |\int_R \Psi(x, \tau) dG(x) - \int_R \Psi(x, \tau) dF_\theta(x)| < \delta, \text{ and}$$

$$\sup_{\tau \in D} |\int_R \frac{\partial}{\partial \tau} \Psi(x, \tau) dG(x) - \int_R \frac{\partial}{\partial \tau} \Psi(x, \tau) dF_\theta(x)| < \delta\}. \qquad (2.17)$$

Now consider the following lemma along the lines of Lemma 3 from Clarke and Futschik (2007).

Lemma 2.2: *Let $A = \{\Psi(x, \tau)|\tau \in D\}$ be a family of real vector functions defined on $\mathcal{E}^k$ and suppose D is compact. Assume Ψ is a continuous function in x and τ. Then A forms an "equicontinuous at each x class of functions."*

Now see condition A_1 of conditions **A** implies the equicontinuity of the classes $\{\Psi(x, \tau)|\tau \in D\}$ and $\{(\frac{\partial}{\partial \tau})\Psi(x, \tau)|\tau \in D\}$. Then, provided the Euclidean norms of these vector and matrix functions are bounded by a continuous function g that is integrable with respect to F_θ, it follows by Theorem 6.2 of Rao (1962) that $f.a.s.l.n.$ the empirical distribution function $F_n \in n(\delta, F_\theta)$. Automatically, then assumption A_4 is satisfied $f.a.s.l.n.$ whenever A_1 and A_2 are satisfied. Thus, we have a corollary to Theorem 2.3.

Corollary 2.1: Assume conditions A_0–A_3 are satisfied on $R \subset \mathcal{E}^k$. Then for the auxiliary selection functional $\rho = |\tau - \theta|$ we find there exists an asymptotically unique consistent solution

$$T[\Psi, \rho, F_n] \to_{a.s.} T[\Psi, \rho, F_\theta] = \theta. \qquad (2.18)$$

□

It can be noted that there are no bounds on Ψ other than the function is bounded above by a continuous and integrable function g, for example see condition A_2. It is the potential unboundedness of the Ψ-function which can lead to non-robustness of the M-estimator, albeit it may be consistent at F_θ. As further explanation of what is meant by an asymptotically unique consistent solution, given any other consistent sequence of solutions $\{\tilde{\theta}_n\}$ of Eq. (2.1) which satisfy $\tilde{\theta}_n \to_{a.s.} \theta$ then it is the case that the event $T[\Psi, \rho, F_n] = \tilde{\theta}_n$ holds for all sufficiently large n.

As we have seen, showing existence of an asymptotically unique consistent root or solution to Eq. (2.1), while somewhat comforting, does not indicate how we should find such a root, and indeed in the case of multiple roots, how we decide which is the consistent estimator. Traditionally, the likelihood approach is to seek the solution of the likelihood that is the global maximum, over all $\theta \in \Theta$, of the likelihood, presuming of course such a maximum exists, and if it does, that it is the unique global maximum. Usually, it is the case that the boundary of the observation space does not depend on θ. A classic example of the opposite would be the data that are uniformly distributed on the open interval $(0, \theta)$. This is not allowed in our discussion. That is, we are assuming suitably regular maximum likelihood surfaces where the global maximum of the likelihood can be found in the interior of the parameter space Θ, by differentiating the likelihood surface to find solutions of maximum likelihood equations. Usually, having found a solution, one checks to see that it is a local maximum by checking the Hessian matrix is negative definite. The choice between two or more solutions is made by choosing the local maximum that gives the largest likelihood, and hopefully having found all the solutions, we identify the unique value that is the largest likelihood value. Clearly, there are parametric families, even in the simple location model of the introduction, where the number of solutions to the likelihood equations increases with increasing sample size. For example, see Reeds (1985) for a description of the asymptotic number of roots of the Cauchy location likelihood equations. Typically, with many parameters, the solution is sought in a region of the parameter space which may be governed by a physical phenomena where the parametric model has an interpretation in the real world. Such regions become more important when the models become more complex, and one thinks of the frontiers opening up in the areas of "Big Data."

In robust M-estimation of location we have seen Hampel's choice of root is that root closest to the median. Using a selection functional such as this, effectively $\rho_o(G, \tau) = |G^{-1}(\frac{1}{2}) - \tau|$ breaks the nexus between the equations and the loss function they minimize, in the case of maximum likelihood the loss function would be the negative of the logarithm of the likelihood for instance. One can choose a solution of the Eq. (2.1) that minimizes a selection functional that need not be related to the equations. The general Theorem 2.4 can also be used to prove consistency of a solution of Eq. (2.1) that minimizes the selection functional over all solutions in the solution set $S(\Psi, F_n)$. For suitable selection functionals, all that is required is that in addition to assumption (2.12) one has

$$\sup_{\tau \in \Theta} |\rho_o(F_n, \tau) - \rho_o(F_\theta, \tau)| \to_{a.s.} 0. \tag{2.19}$$

This is equivalent to assumption (4.1) in Clarke (1990) which is

$$\text{for every } \epsilon > 0 \sup_{\tau \in \Theta} |\rho_o(F_n, \tau) - \rho_o(F_\theta, \tau)| < \epsilon \text{ holds } f.a.s.l.n. \tag{2.20}$$

See that in the case of the selection functional of Hampel (1974) satisfies Eq. (2.12) since clearly for every open neighbourhood of θ,

$$\inf_{\tau \notin N} \rho_o(F_\theta, \tau) - \rho_o(F_\theta, \theta) = \inf_{\tau \notin N} |\theta - \tau| > 0.$$

Moreover, it is well known for empirical processes that at the normal parametric model located at θ that the quantile $F_n^{-1}(\frac{1}{2}) \to_{a.s.} \theta$, the median is consistent for θ, which ensures Eq. (2.19).

Indeed we see now that if we define neighborhoods

$$n^*(\delta, F_o) = \{G \in \mathcal{G}|$$

$$\sup_{\tau \in D} | \int_R \Psi(x, \tau)dG(x) - \int_R \Psi(x, \tau)dF_\theta(x)| < \delta, \text{ and}$$

$$\sup_{\tau \in D} | \int_R \frac{\partial}{\partial \tau}\Psi(x, \tau)dG(x) - \int_R \frac{\partial}{\partial \tau}\Psi(x, \tau)dF_\theta(x)| < \delta, \text{ and}$$

$$\sup_{\tau \in \Theta} |\rho_o(G, \tau) - \rho_o(F_\theta, \tau)| < \delta\}, \tag{2.21}$$

then we have the following corollary to Theorems 2.3 and 2.4.

Corollary 2.2: Assume conditions A_0–A_3 are satisfied on $R \subset \mathcal{E}^k$. Assume the selection functional ρ_o satisfies Eqs. (2.12) and (2.19). Then

$$T[\Psi, \rho_o, F_n] \to_{a.s.} T[\Psi, \rho_o, F_\theta] = \theta.$$

□

This simply follows since under the assumptions of the corollary for every $\delta > 0$, $F_n \in n^*(\delta, F_\theta) f.a.s.l.n.$ To illustrate why the corollary follows consider the following illustration of Figure 2.1 first published in Clarke (1990). This shows us the argument for univariate parameters θ. The multivariate extension can be carried out similarly using multivariate parameter balls instead of intervals on the real line. Proceeding with the argument, by Corollary 2.1 we know there exists a $\kappa > 0$ such that a root $\theta_n^* = T[\psi, \rho, F_n]$ of Eq. (2.1), noting that for univariate parameters $\Psi \equiv \psi$ in this case, exists and is unique in $(\theta - \kappa, \theta + \kappa)$ for all sufficiently large n. Moreover, given arbitrary $0 < \kappa^* < \kappa$, $\theta_n^* \in (\theta - \kappa^*, \theta + \kappa^*) f.a.s.l.n$. Define

$$\delta(\kappa) = \inf_{\tau \notin (\theta - \kappa, \theta + \kappa)} \rho_o(F_\theta, \tau) - \rho_o(F_\theta, \theta) > 0. \tag{2.22}$$

By continuity choose $0 < \kappa^* < \kappa$ so that

$$\sup_{\tau \in (\theta - \kappa^*, \theta + \kappa^*)} |\rho_o(F_\theta, \tau) - \rho_o(F_\theta, \theta)| < \frac{\delta(\kappa)}{2}. \tag{2.23}$$

Note that any other root $\tilde{\theta}_n$ of Eq. (2.1), if it exists, lies outside of $(\theta - \kappa, \theta + \kappa) f.a.s.l.n$. Setting $\epsilon = \delta(\kappa)/4$ in Eq. (2.20) we obtain for such a root

$$\rho_o(F_n, \tilde{\theta}_n) > \rho_o(F_\theta, \tilde{\theta}_n) - \frac{1}{4}\delta(\kappa) f.a.s.l.n \qquad \text{by} \qquad (2.20)$$

$$> \rho_o(F_\theta, \theta) + \frac{3}{4}\delta(\kappa) \qquad \text{by} \qquad (2.22)$$

$$> \sup_{\tau \in (\theta - \kappa^*, \theta + \kappa^*)} \rho_o(F_\theta, \tau) + \frac{1}{2}\delta(\kappa) \quad \text{by} \qquad (2.23)$$

$$> \rho_o(F_n, \theta_n^*) f.a.s.l.n$$

Hence, it follows that $\theta_n^* = T[\psi, \rho_o, F_n] f.a.s.l.n$ because of the definition (2.3).

Naturally, if there is only one solution of the equations then there is no need to employ the selection functional, and consistency follows from Corollary 2.1. One can imagine though that, for example, if one's likelihood surface was to have several hills and valleys and one needed to assert global consistency, then a proof can be arranged according to Corollary 2.2 by showing say Eq. (2.19) whenever $\rho_o(G, \tau) = -\int_R \log f_\tau(x) dG(x)$, since by the arguments of Wald (1949) Eq. (2.12) follows, at least for identifiable parametric families with regular densities $f_\tau(x)$. However, for most parametric families with exponentially decreasing tails it will follow that the MLE is not robust. Correspondingly we do not pursue the arguments for consistency and asymptotic normality here since they do not become the real objective of this book.

On the other hand, a key concept underlying robustness is that of weak continuity of the estimating functional.

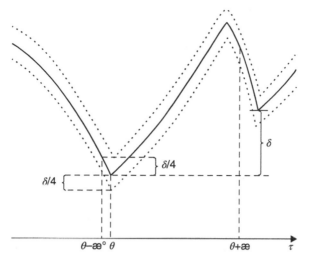

FIGURE 2.1 $\rho_o(F_\theta, \tau)$, ..., $\rho_o(F_\theta, \tau) \pm \delta(\kappa)/4$ for an imaginary selection functional. (*Source*: Clarke (1990). Reproduced with the permission of University of Wroclaw.)

2.3 WEAK CONTINUITY AND WEAK CONVERGENCE

Weak convergence is best articulated through the following proposition regarding convergence of measures on $(R, \mathcal{B})$, see Alexandroff (1943) or Billingsley (1956):

Proposition 2.1: The following statements are equivalent:

(1) $\mu_n \Rightarrow \mu$
(2) $\lim_{n \to \infty} \mu_n(A)$ for each continuity set A of μ and
(3) for each bounded and uniformly continuous function $h(x)$ on R

$$\lim_{n \to \infty} \int h \, d\mu_n = \int h \, d\mu.$$

$\square$

We say a sequence of random variables X_n converges in distribution to a random variable X if $P_n \Rightarrow P$, where P_n and P are the distributions of X_n and X respectively. Further discussion on this matter is given in Billingsley (2009, p. 25). A result of Prohorov (1956) is that if $\{\mu_n\}$ is any sequence of measures such that $\mu_n \Rightarrow \mu$ then this is equivalent to $d_p(\mu_n, \mu) \to 0$. We now include the following formal definition.

Definition 2.1: Let T be defined everywhere in $\mathcal{G}$. Then T is defined to be weakly continuous at $F \in \mathcal{G}$ if for every $\epsilon > 0$ there exists $\delta > 0$ such that for every $G \in \mathcal{G}$, $d_p(F, G) < \delta$ implies $|T[F] - T[G]| < \epsilon$.

Theorem 2.3 can be used to show existence of a weakly continuous root of Eq. (2.2) provided conditions **A** hold with respect to the Prohorov neighborhoods. That this follows for Ψ which are uniformly bounded and continuous on $R \times D$ with bounded and continuous partial derivatives on $R \times D$ results from the combination of Lemma 2.2 and Theorem 2.1. See then that for such Ψ conditions A_1, A_2 imply A_4. One only then needs to check Fisher consistency, meaning that $T[\Psi, \rho, F_\theta] = \theta$ and non-singularity of the matrix $M(\theta)$.

In order that one retains the weakly continuous root when one employs a selection functional ρ_o to select from among the roots of the Eq. (2.2) depends on that selection functional satisfying conditions (2.12) and (2.13) in Theorem 2.4. We shall illustrate how one may show this for the selection functionals used in Chapter 1 for location and for location and scale. Consider the following lemma.

Lemma 2.3: *Assume F to be an absolutely continuous distribution with support on $\mathcal{E}$. Let neighborhoods be defined by one of the metrics d_k, d_L, or d_p and let t be fixed. Then for any $\eta > 0$ there exists a neighborhood $n(\epsilon, F)$ for which $\epsilon > 0$ and*

$$\sup_{G \in n(\epsilon, F)} |F^{-1}(t) - G^{-1}(t)| < \eta \quad where \quad G^{-1}(t) = \inf\{y | G(y) \geq t\}.$$

Proof. Define $a = F(F^{-1}(t) - \eta/2) < t$, $b = F(F^{-1}(t) + \eta/2) > t$. Consider neighborhoods defined by d_k. Let $\epsilon = \min(b - t, t - a)/4$. Then

$$\sup_{G \in n(\epsilon, F)} |G(F^{-1}(t) - \frac{\eta}{2}) - a| < \frac{t - a}{4} \Rightarrow G(F^{-1}(t) - \frac{\eta}{2}) < t;$$

$$\sup_{G \in n(\epsilon, F)} |G(F^{-1}(t) + \frac{\eta}{2}) - b| < \frac{b - t}{4} \Rightarrow G(F^{-1}(t) + \frac{\eta}{2}) > t.$$

Hence, $F^{-1}(t) - \eta/2 < G^{-1}(t) \leq F^{-1}(t) + \eta/2$ which proves the result for d_k. Since F is absolutely continuous with support $\mathcal{E}$, it has a bounded density. Suppose the upper bound is $\tilde{c}$. Then it can be shown that (see Problem 2.3)

$$d_k(G, F) \leq (\tilde{c} + 1)d_L(G, F) \leq (\tilde{c} + 1)d_p(G, F). \tag{2.24}$$

Hence, the result holds also for d_L and d_p. $\qquad\qquad\square$

Clearly, then the selection functional $\rho_o(G, \tau) = |G^{-1}(1/2) - \tau|$ as a result of Lemma 2.3 satisfies Eq. (2.13) at the normal location distribution, and it is easy to see that as a result of Theorem 2.4 that the estimator for location governed by the Tukey bisquare of Eq. (1.6) and this selection functional is weakly continuous. For instance, consider families $\{\psi(x, \tau) = \psi_{BS}(x - \tau)|\tau \in \mathcal{E}\}$ which are uniformly bounded and equicontinuous, again with uniformly bounded and equicontinuous families of partial derivatives.

For location and scale parameters estimated simultaneously, the selection functional of Eq. (1.23), written

$$\rho(G; \tau_1, \tau_2) = (G^{-1}(1/2) - \tau_1)^2 + \left(\frac{G^{-1}(3/4) - G^{-1}(1/4)}{\Phi^{-1}(3/4) - \Phi^{-1}(1/4)} - \tau_2 \right)^2,$$

satisfies Eq. (2.13) at the normal distribution and may be employed with more general redescending Ψ.

More generally under the conditions of Theorem 2.4, we see that the resulting M-functional $T[\Psi, \rho_o, \cdot]$ is weakly continuous at F_θ. Following the results of Theorem 1 in Hampel (1971) [see also Hampel et al. (1986)] regarding the robustness of $T_n(X_1, \ldots, X_n) = T[F_n]$, we can only infer robustness at the distribution F_θ.

On the other hand, an even broader result is argued in Clarke (2000a), where it is shown there is a Prohorov neighborhood $U(F_\theta)$ of F_θ where $T[\Psi, \rho_o, \cdot]$ is continuous at each $G \in U(F_\theta)$. That is, we may apply Theorem 2 in Hampel (1971), obtaining the qualitative robustness of T_n at G, for all G in a neighborhood $U(F_\theta)$, together with the consistency of T_n to $T[G]$.

We list again the specific conditions and state the result of Theorem 1 of Clarke (2000a) and its corollary.

Conditions W

$W_0 \equiv A_0$

$W_1 \equiv A_1$

W_2 $\{\Psi(x, \tau)|\tau \in D\}$, $\{\frac{\partial}{\partial \tau}\Psi(x, \tau)|\tau \in D\}$ are bounded above in Euclidean norm by a finite constant.

$W_3 \equiv A_3$

Theorem 2.5: *(Clarke, 2000a, Theorem 1). Assuming conditions W, there exists a Prohorov neighborhood $U(F_\theta)$ of F_θ such that the functional defined via $T[\Psi, \rho, \cdot]$ is weakly continuous at each $G \in U(F_\theta)$.*

Corollary 2.3: Assuming conditions **W**, there exists a qualitatively robust M-functional estimator T_n which is robust and consistent tending to $T[G]$, for all $G \in U(F_\theta)$.   □

The relationship between weak continuity and consistency can be seen to follow from Prohorov's result, continuity of the functional $T[\Psi, \rho_o, \cdot]$ and a result of Varadarajan (1958) which states with probability one $F_n \Rightarrow G$, whence continuity gives $T[\Psi, \rho_o, F_n] \to_{a.s.} T[\Psi, \rho_o, G]$. Note clearly functionals $T[\Psi, \cdot]$ with unbounded Ψ-functions in the x variable are not continuous with respect to the Prohorov metric and yet they can be strongly consistent at the model F_θ. This then is a fundamental delineation of non-robust M-functionals from robust M-functionals.

While clearly unboundedness of Ψ is one way an M-functional may not be weakly continuous in a neighborhood of F_θ, there are also potentially going to be issues if $\Psi(x, \tau)$ is a noncontinuous function, not the least of which is the existence of a solution of Eq. (2.1).

2.4   FRÉCHET DIFFERENTIABILITY

The Fréchet derivative was introduced into the statistics literature by Kallianpur and Rao (1955). Their reason for using the derivative due to Fréchet (1925) was that it could potentially put any estimator, but in particular the MLE, in the same framework by which asymptotic variance of the estimators could be compared. According to Kallianpur and Rao (1955), Fisher (1925) postulated that in regard to the MLE, "It leads to efficient (asymptotically minimum variance) estimates in the class of asymptotically normal estimates (1925, p. 711)." Kallianpur and Rao (1955) in a sense were attempting to give some justification for this statement. Also while the lower bound for variance of unbiased estimates is now known as the Cramér–Rao lower bound, the class of asymptotically normal estimates is much broader, and Kallianpur and Rao were the first in wanting to quantify in some way such a class to give credence to Fisher's notion of efficiency.

To consider the meaning of Fréchet derivative in a useful manner for statistical functionals, restrictions must be put on the domain of the functional and therefore also the derivative. Let the linear space spanned by the differences $F - G$ of members of $\mathcal{G}$ be denoted by $\mathcal{D}$. The real vector functional T is defined on $\mathcal{G}$ and d is a metric on $\mathcal{G}$.

Definition 2.2:   (*The Fréchet Derivative*)
The statistical functional T is said to be Fréchet differentiable at $G \in \mathcal{G}$ with respect to the pair $(\mathcal{G}, d)$ when it can be approximated by a linear functional $T'_G(\cdot)$ defined on $\mathcal{D}$, in the sense that

$$|T[F] - T[G] - T'_G(F - G)| = o(d(F, G)) \tag{2.25}$$

as $d(F, G) \to 0, F \in \mathcal{G}$. The functional $T'_G(\cdot)$ is known as the Fréchet derivative of the functional T at G.

As noted above, the distance metric that typifies small departures in both rounding errors and fatter tails is the Prohorov metric, and continuity of the functional

$T[\cdot]$ with respect to this metric has automatic implications with respect to consistency of the functional in neighborhoods of the distribution of interest, F_θ say, whereas say the Kolmogorov metric can be more useful in say proofs of asymptotic normality using the Fréchet expansion which we shall see later.

Theorem 2.6: *(Clarke, 1983a, Theorem 5.1) Let $\rho(G, \tau) = |\tau - \theta|$ and assume conditions A hold with respect to this functional and neighborhoods generated by the metric d on $\mathcal{G}$. Suppose for all $G \in \mathcal{G}$*

$$\int_R \Psi(x, \theta)\mathrm{d}(G - F_\theta)(x) = O(d(G, F_\theta)). \tag{2.26}$$

Then $T[\Psi, \rho, \cdot]$ is Fréchet differentiable at F_θ with respect to $\mathcal{G}, d$, and has derivative

$$T'_{F_\theta}(G - F_\theta) = -M(\theta)^{-1}\int_R \Psi(x, \theta)\mathrm{d}(G - F_\theta)(x). \tag{2.27}$$

Proof. Abbreviate $T[\Psi, \rho, \cdot] = T[\cdot]$ and let κ^*, ϵ be given by Theorem 2.3. Let $\{\epsilon_k\}$ be so that $\epsilon_k \downarrow 0^+$ as $k \to \infty$ and let $\{G_k\}$ be any sequence such that $G_k \in n(\epsilon_k, F_\theta)$. Note that $n(\epsilon_k, F_\theta)$ is the set of distributions within distance ϵ_k of F_θ. It is sufficient to show

$$||T[G_k] - T[F_\theta] - T'_{F_\theta}(G_k - F_\theta)|| = o(\epsilon_k).$$

By Theorem 2.3, $T[G_k]$ exists and is unique in $U_{\kappa^*}(\theta)$ for $k > k_o$ where $\epsilon_{k_o} \leq \epsilon$. By A_4 see that

$$||M(\tau, G) - M(\tau, F_\theta)|| \to_{k \to \infty} 0 \text{ uniformly in } \tau \in D. \tag{2.28}$$

Consider the two-term expansion (mean value theorem),

$$0 = K_{G_k}(T[G_k]) = K_{G_k}(\theta) + M(\tilde{\tau}_k, G_k)(T[G_k] - \theta), \tag{2.29}$$

where $||\tilde{\tau}_k - \theta|| \leq ||T[G_k] - \theta||$ which tends to zero as $k \to \infty$ by Theorem 2.3, and $\tilde{\tau}_k$ is evaluated at different points for each component function expansion (i.e. takes different values in each row of M). See from Eqs. (2.29) and (2.26) that

$$||T[G_k] - \theta|| = O(K_{G_k}(\theta)) = O(\epsilon_k).$$

Also

$$T[G_k] - \theta = -M(\theta)^{-1}K_{G_k}(\theta)$$
$$+ M(\theta)^{-1}\{M(\tilde{\tau}_k, G_k) - M(\theta)\}(T[G_k] - \theta). \tag{2.30}$$

By continuity of $M(\tau, F_\theta)$ in τ and Eq. (2.28),

$$||M(\tilde{\tau}_k, G_k) - M(\theta)|| = o(1).$$

So

$$||T[G_k] - \theta - T'_{F_\theta}(G_k - F_\theta)|| = o(1)O(d(G_k, F_\theta)) = o(\epsilon_k).$$

□

Corollary 2.4: If conditions of Theorem 2.6 hold and ρ_o satisfies conditions of Theorem 2.4, then the functional $T[\Psi, \rho_o, \cdot]$ is Fréchet differentiable at F_θ with respect to $(\mathcal{G}, d)$ having the same derivative (2.27).   □

The following consequence of Fréchet differentiability is noted in Huber (1981) and also Boos and Serfling (1980) (Lemma 1.1).

Corollary 2.5: Assume $R = \mathcal{E}^k$, and assume $T[\Psi, \rho_o, \cdot]$ is Fréchet differentiable at F_θ with respect to $(\mathcal{G}, d_k)$, then $\sqrt{n}(T[\Psi, \rho_o, F_n] - T[\Psi, \rho_o, F_\theta])$ converges in distribution to a multivariate normal random variable with mean zero and variance covariance matrix, given by

$$\Sigma(T, F_\theta) = M(\theta)^{-1} \int_R \Psi(x, \theta)\Psi(x, \theta)^T dF_\theta(x)\{M(\theta)^{-1}\}^T \qquad (2.31)$$

where integration is carried out componentwise.   □

For a description of the proof of this result, we see that from the Fréchet expansion

$$\sqrt{n}(T[\Psi, \rho_o, F_n] - T[\Psi, \rho_o, F_\theta]) \approx M(\theta)^{-1} \frac{1}{\sqrt{n}} \sum \Psi(X_i, \theta)$$

$$+ o(\sqrt{n} d_k(F_n, F_\theta)), \qquad (2.32)$$

The normed sum of random variables where the function $\Psi(X, \theta)$ has zero expectation at the model F_θ, as required by assumption A_0 of Fisher consistency, converges in distribution by the multivariate central limit theorem of Rao(1973, p. 128) to the aforesaid distribution of the Corollary 2.5. By the multivariate version of the Dvoretzky–Kiefer–Wolfowitz inequality, which is discussed in Kiefer (1961), we have that $d_k(F_n, F_\theta) = o_p(1/\sqrt{n})$, whence we may apply Slutsky's theorem to infer the result.

Getting a proof of asymptotic normality of the estimator $T[\Psi, \rho_o, \cdot]$ via the vehicle of Fréchet differentiability at the model distribution, eluded the original authors Kallianpur and Rao (1955) in the sense that while Rao (1957) was able to show Fréchet differentiability of the MLE in the multinomial distribution, Kallianpur (1963) reported that in general neither author could under any reasonable set of assumptions (on the density function in the continuous, and the probability function in the infinite discrete case) prove Fréchet differentiability of the MLE. The reason for this difficulty soon becomes apparent when one examines the efficient score function (2.5) for many popular parametric families including the normal, exponential, Poisson, etc. where we find the corresponding Ψ-function is unbounded in the observation space variable. Recall the discussion following Theorem 2.1 suggests that one requires boundedness of the Ψ-function in the x variable if say one is wanting to establish Fréchet differentiability with respect to the Kolmogorov distance metric.

Example 2.1: An example of a Fréchet differentiable MLE is afforded by the Cauchy distribution location and scale parametric family. Consider the parametric family governed by density $\sigma[\pi\{\sigma^2 + (x - \mu)^2\}]^{-1}$, and assume $-\infty < \mu < \infty$, and $\sigma > 0$ so that the parameter $\theta = (\mu, \sigma)^T$ is unknown. The MLE is a solution of Eq. (1.20) with $\Psi = (\psi_1, \psi_2)^T$ and $\tau = (\tau_1, \tau_2)^T$, with

$$\psi_1(x, \tau) = -\frac{x - \tau_1}{\tau_2^2 + (x - \tau_1)^2} \quad , \quad \psi_2(x, \tau) = 1 - \frac{2(x - \tau_1)^2}{\tau_2^2 + (x - \tau_1)^2}. \quad (2.33)$$

Restricting the parameter space $\Theta = \{-\infty_1 < \tau_1 < \infty, \quad \eta < \tau_2 < \infty\}$ for some small positive η ensures Ψ and its partial derivatives with respect to τ are uniformly bounded. For any compact set D containing θ in its interior conditions A_0, A_1, A_2, A_3 hold with respect to both the Kolmogorov metric and the Prohorov metric and as argued previously bounded Ψ implies conditions W are met implying Theorem 2.5 holds. That is the MLE is robust in the sense of Hampel (1971) Theorem 2. Moreover, since Ψ is a function of total bounded variation, then by integration by parts

$$\int_{\mathcal{E}} \Psi(x, \theta) d(G - F_\theta) = -\int_{\mathcal{E}} (G - F_\theta)(x) d\Psi(x, \theta)$$

which implies Eq. (2.26) holds with respect to the Kolmogorov metric. Also for bounded Ψ with bounded partial derivatives on $\mathcal{E} \times D$ conditions A_1, A_2 imply A_4. Hence by Theorem 2.6 there exists a Fréchet differentiable root of Eq. (2.2) at F_θ with respect to the Kolmogorov metric. It is noted by Copas (1975) that the Cauchy joint likelihood for both location and scale parameters is unimodal with one stationary point. Hence the MLE corresponds also to the asymptotically efficient root in the sense of Kallianpur and Rao (1955). □

In a similar vein one can also show that the Tukey bisquare estimator for location, with a choice of root closest to the median, is weakly continuous in an open neighborhood of the normal model and Fréchet differentiable with respect to the Kolmogorov metric at the normal model.

2.5   THE INFLUENCE FUNCTION

A discussion of the functional approach to estimation necessarily must include a discussion of the influence function, which following Kallianpur and Rao (1955) was introduced in Hampel (1968, 1974) and led to the wide ranging book by Hampel et al. (1986). Whereas the Fréchet derivative for many estimators (though not M-estimators with suitably bounded and continuous Ψ and classes of L-estimators with suitable weight functions (see Section 7.1.1), is somewhat elusive, the influence function requires no appeal to bothersome regularity conditions. The authors of Hampel et al. (1986, p. 84) conclude that "This makes the range of its applicability very large, as it can be calculated in all realistic situations." Briefly, letting $x \in R$ and defining δ_x to be that distribution in $\mathcal{G}$ that attributes point one mass of the distribution to the point x, and considering the ϵ-contaminated distribution

$$F_{\theta,x,\epsilon}(y) = (1 - \epsilon)F_\theta(y) + \epsilon\delta_x(y)$$

the infinitesimal influence of a single point $x \in R$ at the distribution F_θ is defined to be

$$\mathrm{IF}(x; F_\theta, T) = \lim_{\epsilon \downarrow 0} \frac{T[F_{\theta,x,\epsilon}] - T[F_\theta]}{\epsilon}. \tag{2.34}$$

Another way to define the influence function is to simply see that

$$\mathrm{IF}(x; F_\theta, T) = \frac{\partial}{\partial \epsilon} T[F_{\theta,x,\epsilon}]|_{\epsilon=0}$$

which is widely exploited in many works, including that of Staudte and Sheather (1990) for example.

From Huber (1996, p. 14) and Hampel et al. (1986, p. 230), we see that the M-estimator has an influence function

$$\mathrm{IF}(x; F_\theta, T) = -M(\theta)^{-1}\Psi(x, \theta) \tag{2.35}$$

To quote Huber "let us remember that the influence function of an M-estimate is simply proportional to Ψ." Recall from Chapter 1 that the influence function ψ_{BS} is bounded, whereas at the standard normal distribution the mean functional is given by the choice of $\psi(x) = x$ which suggests that an individual observation can have unbounded influence on the estimate. Thus, for an M-estimator having a bounded and smooth Ψ can lead to both Fréchet differentiability of the M-functional and also a bound on the influence of any point in the observation space on the estimate. We are assuming of course that assumptions A_0 and A_3 are also satisfied.

To quantify this further, Hampel (1974) and Hampel et al. (1986) introduced the gross error sensitivity, whereby if one considers the gross error model $\mathcal{F} = \{F|F = (1 - \epsilon)F_\theta + \epsilon H, H \in \mathcal{G}\}$, we have as a result of Fréchet differentiability

$$T[F] - T[F_\theta] = \int_R \text{IF}(x; F_\theta, T)dF + o(\epsilon)$$

$$= \epsilon \int_R \text{IF}(x; F_\theta, T)dH(x) + o(\epsilon),$$

which gives the worst that nature can do to bias your statistical estimator as

$$\sup_{H \in \mathcal{G}} |T[F] - T[F_0]| \leq \epsilon \sup_{H \in \mathcal{G}} |\int_R \text{IF}(x; F_\theta, T)dH(x)| + |o(\epsilon)|$$

$$\leq \epsilon \sup_{H \in \mathcal{G}} \int_R |\text{IF}(x; F_\theta, T)|dH(x) + |o(\epsilon)|$$

$$\leq \epsilon \cdot \gamma^* + |o(\epsilon)|$$

where

$$\gamma^* = \sup_{x \in R} |\text{IF}(x; F_\theta, T)|. \tag{2.36}$$

Hampel refers to γ^* as the *gross-error sensitivity*, and Hampel et al. (1986) make much on optimizing the asymptotic variance of a statistic subject to a bound on the gross-error sensitivity. Huber (1996) in a similar discussion to the above notes that if one strays from Fréchet differentiability, it is possible to find estimators where $\gamma^* < \infty$, but the asymptotic bias is infinite, and another case where $\gamma^* = \infty$, but the limit of the asymptotic bias is zero as $\epsilon \to 0$. Notably, these are not M-estimators.

Returning now to the Fréchet derivative of an M-functional, see from Eqs. (2.27) and (2.35)

$$T'_{F_\theta}(G - F_\theta) = \int_R \text{IF}(x; F_\theta, T)d(G - F_\theta)(x)$$

and moreover it can be assumed without loss of generality, by Fisher consistency

$$\int_R \text{IF}(x; F_\theta, T)dF_\theta = 0.$$

Subsequently, the asymptotic variance for a Fréchet differentiable M-functional is

$$\Sigma(T, F_\theta) = \int_R \text{IF}(x; F_\theta, T)\text{IF}(x; F_\theta, T)^T dF_\theta(x)$$

The important argument raised by Kallianpur and Rao (1955) which is also described in Huber (1996, p. 23; Hampel et al. 1986, p. 86) is the following. We give this in the case of a univariate parameter and assume that the estimator functional is Fisher consistent and satisfies $T[F_\theta] = \theta$ and is Fréchet differentiable. Then quoting with permission from Huber (1996, p. 23) T is "asymptotically efficient if and only if its influence function satisfies

$$\mathrm{IF}(x; F_\theta, T) = \frac{1}{\mathcal{I}(F_\theta)} \frac{\partial}{\partial \theta}(\log f_\theta(x))$$

Here, f_θ is the density of F_θ, and

$$\mathcal{I}(F_\theta) = \int (\frac{\partial}{\partial \theta} \log f_\theta(x))^2 dF_\theta(x) \qquad (2.37)$$

is the Fisher information." Huber (1996) assumes "$d_L(F_\theta, F_{\theta+\delta}) = O(\delta)$, that

$$\frac{f_{\theta+\delta} - f_\theta}{\delta \cdot f_\theta} \xrightarrow{L_2(F_\theta)} \frac{\partial}{\partial \theta} \log f_\theta$$

and that

$$0 < \mathcal{I}(F_\theta) < \infty.$$

Then by the definition of the Fréchet derivative

$$T[F_{\theta+\delta}] - T[F_\theta] - \int_R \mathrm{IF}(x; F_\theta, T)(f_{\theta+\delta} - f_\theta)dx = o(d_L(F_\theta, F_{\theta+\delta})) = o(\delta)$$

We divide this by δ and let $\delta \to 0$. This gives

$$\int_R \mathrm{IF}(x; F_\theta, T)\{\frac{\partial}{\partial \theta} \log f_\theta(x)\} f_\theta(x)dx$$

The Schwarz inequality applied to this equation first gives that the asymptotic variance $\sigma^2(F_\theta, T)$ of $\sqrt{n}T_n$ satisfies

$$\sigma^2(F_\theta, T) = \int_R \mathrm{IF}(x; F_\theta, T)^2 dF_\theta(x) \geq \frac{1}{\mathcal{I}(F_\theta)}$$

and second, that we can have equality only if $\mathrm{IF}(x; F_\theta, T)$ is proportional to $\frac{\partial}{\partial \theta} \log f_\theta(x)$."

This result then gives in a sense some credence to Fisher's original postulate made in Fisher (1925). For among all Fréchet differentiable estimators, the MLE functional achieves the minimum asymptotic variance, when the MLE itself is Fréchet differentiable. Examples of this include location parametric families for the Student's t-distributions, where the number of degrees of freedom is assumed finite. The classic example already mentioned above is the Student's t-distribution with one degree of freedom being the Cauchy distribution. Rao (1957) shows that the MLE of a multinomial distribution is Fréchet differentiable. Unfortunately, the requirement of a continuous bounded efficient score functional which would lead to a bounded continuous influence function is, as has been noted previously, not available for many of the most important parametric families, including the normal, the exponential, and the Poisson distributions for example, and we shall see later that this is also the case for finite mixtures of these distributions. One can at the parametric model find the asymptotic limit distribution with asymptotic variance being the inverse of Fisher information using say conditions **A** with appropriate neighborhoods assuming appropriate regularity conditions on the parametric densities. See ahead in Section 3.4 for example. However, unlike the Cauchy, or Student's t-distributions we find $\gamma^* = +\infty$, and the estimators are sensitive to departures in the tails of those distributions.

2.6   EFFICIENCY FOR MULTIVARIATE PARAMETERS

The asymptotic efficiency of the maximum likelihood statistic carries over easily to vector parameters. For example, consider a parameter space $\Theta \subset \mathcal{E}^r$ and any statistic

$$t = l^T \theta = l_1 \theta_1 + \ldots + l_r \theta_r.$$

Suppose $T[\cdot]$ to be a Fisher consistent Fréchet differentiable functional at F_θ. Then it follows that the statistic $T_l[\cdot] = l^T T[\cdot]$ is a Fisher consistent Fréchet differentiable statistic for t and moreover the asymptotic variance of $T_l[\cdot]$ is given by

$$\sigma^2(F_\theta, T_l) = l^T \sigma(T, F_\theta) l.$$

Assume now that the MLE of θ, call it $T_{\text{MLE}}[\cdot]$, is Fréchet differentiable, whence it has asymptotic variance $\mathcal{I}(\theta)^{-1}$ at the model F_θ. By the invariance principle of the MLE, the MLE of any function of θ is the function of the MLE of θ. See for example Casella and Berger (2002, Theorem7.2.10, pp. 320–321) including the discussion. Hence, it follows the MLE of $l^T \theta$ is $l^T T_{\text{MLE}}[F_n]$ which has asymptotic variance $l^T \mathcal{I}(\theta)^{-1} l$ and it follows from the previous section that

$$l^T \Sigma(T, F_\theta) l \geq l^T \mathcal{I}(\theta)^{-1} l, \qquad (2.38)$$

and this is true for all l. That is,

$$l^T (\Sigma(T, F_\theta) - \mathcal{I}(\theta)^{-1}) l \geq 0 \qquad (2.39)$$

with equality if T is the MLE. That is, the matrix $\Sigma(T, F_\theta) - I(\theta)^{-1}$ is nonnegative definite. Arguments can be generalized to parameters with more dimensions and indeed in the spirit of Kallianpur and Rao (1955), to consider estimating functionals $T = (T_1, T_2, \dots)$ which are Fisher consistent for the vector of parametric functions $(\Phi_1, \Phi_2, \dots)$. Details are in section 4 of that paper.

2.7   OTHER APPROACHES

It is interesting to note the historical rise and fall and rise again of the Fréchet derivative in statistics. Following the remarks of Kallianpur (1963), the Fréchet derivative did not take root in the statistics literature until its importance was recognized in the area of robustness, where the functional approach came again to the fore. Fréchet differentiability was resurrected by Huber (1977) and the PhD thesis by Reeds (1976), which celebrated also the weaker Hadamard or compact derivative. In fact there appears to be three main forms of functional derivative. To briefly discuss these, we consider extending the space $\mathcal{G}$ to a linear space of distributions corresponding to finite signed measures

$$\mathcal{M} = \{aF + bG : a, b \text{ real}; F, G \in \mathcal{G}\},$$

$\mathcal{M}$ is a normed linear space with respect to $||H|| = \sup_x |H(x) - H(-\infty)|$, for example. The extension of the domain of functionals from $\mathcal{G}$ to $\mathcal{M}$ here is only for the purposes of discussion of the derivatives of functionals. Rather than considering just a map $T : \mathcal{G} \to \Theta \subset \mathcal{E}^r$ we denote the map $T : \mathcal{M} \to \mathcal{N}$, where $\mathcal{N}$ is a normed linear space. For instance, define derivatives for T at $F \in \mathcal{M}$ as follows. We define a *remainder* by letting $H \in \mathcal{M}$ and a continuous linear map $\text{Lin} : \mathcal{M} \to \mathcal{N}$ so that

$$R(G + tH) = \begin{cases} T[G + tH] - T[G] - \text{Lin}[tH], & t \neq 0 \\ 0 & t = 0 \end{cases} \tag{2.40}$$

Suppose

$$\frac{R(G + tH)}{t} \to 0 \text{ in the norm on } \mathcal{N} \text{ as } t \to 0.$$

(i) T is *Gâteaux differentiable* at G with derivative $T'_G = \text{Lin}$ if Eq. (2.40) holds for all $H \in \mathcal{M}$.

(ii) T is *Hadamard differentiable* (or compact differentiable) if (2.40) holds uniformly for H lying in an arbitrary compact set of $\mathcal{M}$; and T'_G is called the *compact derivative* of T at G.

(iii) T is *Fréchet differentiable* if Eq. (2.40) holds uniformly for H lying in an arbitrary bounded subset of $\mathcal{M}$; and T'_G is called the *Fréchet derivative* of T at G.

The derivatives are from weaker to stronger. The weakest derivative is the Gâteaux derivative. According to Huber (1977) and Hampel et al. (1986), a functional T is Gâteaux differentiable at the distribution G in $\mathcal{G}$, if there exists a real functional ψ such that for all $F \in \mathcal{G}$ it holds that

$$\lim_{t \to 0} \frac{T[(1-t)G + tF] - T[G]}{t} = \int \psi(x) dF(x)$$

which may also be written as

$$\frac{\partial}{\partial t}(T[(1-t)G + tF])|_{t=0} = \int \psi(x) dF(x). \tag{2.41}$$

Putting $F = G$ in Eq. (2.41) it becomes clear that

$$\int \psi(x) dG(x) = 0$$

so one can replace $dF(x)$ by $d(F - G)(x)$ in Eq. (2.41). Further, by substituting $F = \delta_x$ it becomes clear that $\psi(x)$ in Eq. (2.41) is in fact $\mathrm{IF}(x; G, T)$ from which one can then explore Taylor type series

$$T[F] = T[G] + \int \mathrm{IF}(x; G, T) d(F - G)(x) + \text{remainder}. \tag{2.42}$$

The compact derivative is in between the Gâteaux derivative and the Fréchet derivative. The Fréchet derivative is too strong for some functionals. A typical example that is provided by Fernholz (1983, Example 2.3.2, p. 14) and Staudte and Sheather (1990, Example 4, p. 303). They show that the median is not Fréchet differentiable at the uniform distribution on (0, 1); however, quantiles are compactly differentiable [see Reeds (1976) and Fernholz (1983)]. The key ingredient that makes the functional not Fréchet differentiable here is that the median, and indeed quantiles, have influence functions that are discontinuous. For example, consider the qth quantile functional $T[F] = F^{-1}(q)$, where F is a distribution on the real line having a density f which is continuous and positive at $x_q = F^{-1}(q)$. Then the influence function of T at F is

$$\mathrm{IF}(x; F, T) = \begin{cases} \frac{(q-1)}{f(x_q)}, & x < x_q \\ 0, & x = x_q \\ \frac{q}{f(x_q)}, & x > x_q \end{cases}. \tag{2.43}$$

See Staudte and Sheather (1990, p. 59) for guidance on its derivation.

As we have seen from the discussion following Corollary 2.5, M-functionals with Ψ-functions that are unbounded in the observation space variable are not Fréchet differentiable with respect to the supremum norm. They have infinite gross error sensitivity. Typical maximum likelihood estimating functionals for many parametric families that have exponentially decreasing tails have unbounded score functions and even though they are regular they are not Fréchet differentiable.

Huber (1981) while noting the advantages of Fréchet differentiability wrote "... Unfortunately the concept of Fréchet differentiability appears to be too strong: in too many cases, the Fréchet derivative does not exist, and even if it does, the fact is difficult to establish." At least for M-estimators with bounded Ψ-functions and under reasonable continuity conditions this is indeed not the case as evidenced by Clarke (1983a, 1986b). At about the same time, Fernholz (1983) produced a monograph which among other results pointed out, as has been noted above, that the median, a well-known robust estimator, is not Fréchet differentiable, but it is compactly differentiable at the normal distribution. These works have spawned numerous articles in different directions. For Gill and Heesterman (1992) show that for most regular parametric familis, the MLE is compactly differentiable, whereupon it is clear that non-robust estimators, such as the MLE for parameters in normal distributions, are also compactly differentiable.

In a work by Bednarski et al. (1991), it is shown that estimators that satisfy certain asymptotic expansions in root n shrinking neighborhoods are asymptotically equivalent to Fréchet differentiable M-estimators with continuous and bounded Ψ-functions. The authors also exhibit an example of the median, which has a non-continuous ψ-function, where the distribution of the median is different to that of the first term in an asymptotic expansion under a root n shrinking regime. (See Lemma 3.2 in Chapter 3 for example.)

Finally, it should be noted that review papers by Gill et al. (1989) and Clarke (2000b) and the book by Rieder (1994), as well as works by van der Vaart (1991) and van der Vaart and Wellner (1996) show the area of differentiable estimating functionals has had an impact on the statistical literature.

PROBLEMS

2.1. A sequence of constants a_n is $o(1)$ so long as $\lim_{n\to\infty} a_n = 0$. A sequence of constants $a_n = o(n^{-k})$ provided $\lim_{n\to\infty} \frac{a_n}{n^{-k}} = 0$. A sequence $\{a_n\}_{n=1}^{\infty}$ is said to be of large order $O(t_n)$ whenever $\{a_n/t_n\}$ is bounded and of small order $o(t_n)$ whenever $\frac{a_n}{t_n} \to 0$ as $n \to \infty$.

(i) Show $a_n = 2n + 1 = O(n)$.

(ii) Show $a_n = n^{-\frac{1}{2}-\epsilon} = o(n^{-\frac{1}{2}})$ for $\epsilon > 0$.

2.2. Using the definition of the Kolmogorov, Lévy distance metrics show that $d_{\mathrm{L}} \leq d_k$ and $d_{\mathrm{L}} \leq d_{\mathrm{p}}$. No general relation holds between d_k and d_{p}.

2.3. Assuming $F(x)$ is a continuous distribution on $\mathcal{E}$ which has a density bounded above by $\tilde{c}$ so that $\sup_{x \in \mathcal{E}} F(x + \delta) - F(x) < \tilde{c}\delta$ uniformly in

$\delta > 0$. Show that $F(x) \le G(x + \delta) + \delta$ and $G(x) \le F(x + \delta) + \delta$ uniformly in $x \in \mathcal{E}$ implies $\sup_{x \in \mathcal{E}} |G(x) - F(x)| < (\tilde{c} + 1)\delta$. Hence, observe

$$d_k(G, F) \le (\tilde{c} + 1)d_L(G, F) \le (\tilde{c} + 1)d(G, F).$$

2.4. Show that if a functional T is Fréchet differentiable at distribution F with respect to the Lévy distance metric, then it is also Fréchet differentiable at F with respect to the Kolmogorov distance metric as well as the Prohorov distance metric. *Hint: Use formula* (2.25).

2.5. Suppose that the functional T is Fréchet differentiable at the distribution F with respect to the Kolmogorov distance metric and assume F satisfies the conditions of Problem 2.3. Then show that the functional T is Fréchet differentiable at F with respect to both d_L and d_p.

2.6. In the theory of contamination, it is fashionable to consider "ϵ-contaminated" neighborhoods of a particular distribution defined as

$$n_\epsilon(F) = \{(1 - \epsilon)F + \epsilon H \mid H \in \mathcal{G}, \ 0 < \epsilon < 1\}.$$

Neighborhoods can also be generated by distance metrics, for example, the Prohorov neighborhood where

$$n_p(\epsilon, F) = \{G \in \mathcal{G} \mid d_p(F, G) < \epsilon\}.$$

Show $n_\epsilon(F) \subset n_p(\epsilon, F)$.

2.7. Find an expression $T[G]$ such that

$$T[F_n] = \frac{\sum\limits_{i=1}^{n} (X_i - \overline{X})^2}{n}$$

Now suppose G has mean μ and variance σ^2. Derive the influence function of T at G. What does the influence function reveal about the behavior of T as an estimator?

2.8. Suppose $X_1, \ldots, X_n$ are independent and identically distributed with distribution function $F = F(x; \mu)$. Assume $F(x; \mu)$ is differentiable with density $F'(x; \mu) = \frac{\partial}{\partial x} F(x; \mu) = f(x - \mu)$. Assume $f(z)$ is symmetric about $z = 0$, whence the common distribution $F(x; \mu)$ is symmetric about μ. An M-estimator $T_n = T_n(X_1, \ldots, X_n)$ of μ satisfies

$$\sum_{i=1}^{n} \psi(X_i - T_n) = 0.$$

Using a particular choice of ψ given by

$$\psi(z) = \frac{\frac{d}{dz}f(z)}{f(z)},$$

use the general form of the influence function, to write down the influence function of this M-estimator for location. Give the particular formula for the asymptotic variance of the M-estimator for this special case. Can you recognize your estimator as being a familiar estimator?

2.9. (a) Let F be an absolutely continuous distribution function (meaning that it has a derivative in the observation space variable) and assume that F has infinite support with density $f(x) > 0$ for all x. Also $F(F^{-1}(t)) = t$ for all $0 < t < 1$. Define

$$G_\epsilon(x) = (1 - \epsilon)F(x) + \epsilon\delta_z(x),$$

where $\delta_z(x)$ is the distribution attributing mass one to the point z. Show that

$$G_\epsilon^{-1}(t) = \begin{cases} F^{-1}(\frac{t}{1-\epsilon}) & \text{if} & F^{-1}(\frac{t}{1-\epsilon}) < z \\ z & \text{if } F^{-1}(\frac{t-\epsilon}{1-\epsilon}) \leq z \leq F^{-1}(\frac{t}{1-\epsilon}) \\ F^{-1}(\frac{t-\epsilon}{1-\epsilon}) & \text{if} & F^{-1}(\frac{t-\epsilon}{1-\epsilon}) > z \end{cases}$$

Again note that for a general distribution $H(x)$, we define

$$H^{-1}(t) = \inf\{y : H(y) \geq t\}.$$

(b) Using the result of part (a), show from first principles that the influence function for the median functional, given by $T[F] = F^{-1}(1/2)$, is given by

$$\text{IF}(z, F, T) = \frac{1}{2f(F^{-1}(1/2))} \, \text{sgn}(z - F^{-1}(1/2)).$$

Here

$$\text{sgn}(x) = \begin{cases} 1 & \text{if } x > 0 \\ -1 & \text{if } x < 0 \\ 0 & \text{if } x = 0 \end{cases}$$

Hence, what is the asymptotic variance for the median?

2.10. Consider $X_1, \ldots, X_n$ to be independently identically distributed with parametric density

$$f(x - \mu) = \frac{1}{\pi\{1 + (x - \mu)^2\}} \qquad -\infty < x < \infty.$$

(a) Write down the explicit form of the maximum likelihood equations used to find the MLE of μ. (You need not solve the equations.)

(b) i. Define a general form for equations used to define an M-estimator of location of a symmetric distribution and show the equations from part (a) are of this M-estimator form. **Hint:** What is the form for the ψ-function?

   ii. Write down the specific form for the influence function of the MLE of μ using the fact that the estimator is an M-estimator for location and

$$\int_{-\infty}^{+\infty} \frac{1 - x^2}{(1 + x^2)^3} dx = \pi/4.$$

(c) Use calculus to sketch the influence function of (b)(ii).

(d) Using your answers from part (b), calculate the gross error sensitivity of the MLE.

(e) Give reasons why or why not you think the MLE for this parametric family should be both robust and efficient.

2.11. Consider the distribution defined on the positive half line with densities

$$f(x; \sigma, p) = [\sigma\Gamma(p)]^{-1} x^{p-1} \exp(-\frac{x}{\sigma})$$

with $\sigma > 0$ and $p > 0$, where σ is a scale parameter and p characterizes the shape of the distribution. Here $\Gamma(p) = \int_0^\infty x^{p-1} \exp(-x) dx$. Derive the influence function of the MLE for the scale parameter, keeping p fixed. Is the MLE robust here?

2.10 Consider $X_1, \ldots, X_n$ to be independently identically distributed with probability density

3

MORE RESULTS ON DIFFERENTIABILITY

3.1 FURTHER RESULTS ON FRÉCHET DIFFERENTIABILITY

It is not possible to describe results on differentiability of statistical estimators without first describing them in relation to M-estimators, since M-estimators include maximum likelihood estimators and these have been the predominant tool of estimation with the rise of statistics in the twentieth century, and remain so this century. It is pertinent then to give some more history of M-estimation.

3.2 M-ESTIMATORS: THEIR INTRODUCTION

The history of robustness is intimately tied up with the theory of M-estimation. Section 1.5 explored ideas essentially emanating from Tukey (1960). It was soon after the appearance of that paper that there appeared the major seminal work of Huber (1964), which defined robustness as an area of study worthy of future generations of research. It was here that M-estimators of location were first introduced. Hampel (1992) writes a celebrating article of this work, discussing its philosophical implications and importance in statistics.

Because of historical reasons and no less the central limit theorem, the focus of the research is around the normal distribution. Classically, the least squares estimator is a solution to the minimization problem

Robustness Theory and Application, First Edition. Brenton R. Clarke.
© 2018 John Wiley & Sons, Inc. Published 2018 by John Wiley & Sons, Inc.
Companion website: www.wiley.com/go/clarke/robustnesstheoryandapplication

$$\rho(F_n, \tau) = \frac{1}{n} \sum_{i=1}^{n} L(X_i - \tau) = \text{min!,} \tag{3.1}$$

where $L(x) = x^2$ is chosen to be the least squares loss function. Minimizing this function by setting the partial derivative with respect to τ, one essentially arrives after scaling at equations (1.7) with $\psi(x) = x$ and the resultant estimator $T[F_n] = \int x \mathrm{d}F_n(x) = \overline{X}$ the usual sample mean. The derivation of this as the maximum likelihood estimator assuming the normal parametric family is discussed in Clarke (2008, p. 197), for example. The discovery of the method of least squares is also discussed in detail in Stigler (1999) with weight of evidence from the main players of Gauss and Legendre who both introduced it at different times. On the other hand, the median is arrived at by solving Eq. (3.1) using the L_1-loss function; $L_1(x) = |x|$. This function is only piecewise differentiable and the corresponding psi function is $\psi(x) = \text{sign}(x)$. Questions of solving or alluding to a possible solution of equations (1.7) are addressed in Huber (1964), for example. The median corresponds to a maximum likelihood estimator for the Laplace distribution; corresponding to the density $f(x) = \exp(-|x|)/2$. More generally, $L(x) = -\log f(x)$, where f is the assumed density of the untranslated distribution F, corresponds to choosing the maximum likelihood estimator, should this be the true F generating the data. The connection of L to the underlying distribution, however, is not necessary. Having chosen a loss function for whatever choice of F, the true distribution from nature may be some $G \in \{(1 - \epsilon)\Phi + \epsilon H, H \in \mathcal{G}, H \text{ symmetric}\}$ and because of symmetry one may still use Eq. (3.1). The resulting estimator is known as an M-estimator. Correspondingly, the M-estimator is governed by Eq. (1.7) with $\psi(x) \propto L'(x)$. Huber (1964) gives various conditions on L and ψ for consistency and asymptotic normality of such M-estimators of location. By narrowing the focus to M-estimators, the most robust estimator based on a measure of asymptotic variance is in fact governed by

$$L(x) = \begin{cases} x^2/2 & |x| < k, \\ k|x| - \frac{1}{2}k^2 & |x| \geq k, \end{cases} \tag{3.2}$$

with k depending on ϵ via

$$\frac{2\phi(k)}{k} - 2\Phi(-k) = \frac{\epsilon}{1 - \epsilon}. \tag{3.3}$$

See Huber and Ronchetti [2009, p. 75, formula (4.15)]. This results in the M-estimator illustrated by the psi function of Figure 1.12, where $\psi \equiv \psi_1$. It is known as a "minimax solution," effectively optimizing against the worst that can happen over the ϵ-contaminated neighborhood. It is also known as the maximum likelihood estimator for what has been called Huber's least favorable distribution. As explained in Hampel et al. (1986, p. 37), in the formalism of a two-person zero-sum game, Nature chooses a distribution in the neighborhood of the model, the statistician chooses an M-estimator via their choice of ψ. The gain for Nature and loss for the statistician is the asymptotic variance, $V(\psi, G) = \int \psi^2 dG / (\int \psi' dG)^2$. Huber's minimax solution is a saddlepoint in the game. Of course there are limiting cases when $k = \infty$ which is the least squares solution, and taking the limit as $k \to 0$ which gives the median.

Huber's (1964) discussion encompasses many aspects, including non-convex loss L, asymmetric distributions H, Winsorization, and estimation of location and scale simultaneously, eventually culminating in Huber's Proposal 2 introduced earlier in Section 1.6.

Soon after Huber's success came Hampel's (1968) PhD dissertation, which also emanated from the Berkeley school of statisticians. Here Hampel introduced the influence curve, later to be called the *influence function*, defined *gross error sensitivity* (as given in formula (2.36), defined *qualitative robustness* of a sequence of estimators $\{T_n\}$, and introduced what is known as the *breakdown point*. All these are discussed further in works of Hampel (1971, 1974), and the many other robustness books on the market. Hampel's (1968) thesis revived the functional approach to statistics having somewhat been forgotten after the works of von Mises (1947) and Kallianpur and Rao (1955).

Springing from Lemma 5 of Hampel (1968) came another affirmation of the loss function (3.2). Putting a bound on the gross error sensitivity allows construction of M-estimators that are as efficient as possible. For a mathematical discussion of this result, see Theorem 1 of Hampel et al. (1986) and the specialization to the location case at the normal model in their formula (2.49). This yields the same estimator as the minimax solution. As explained in Hampel et al. (1986, p. 44), "A low gross error sensitivity, however, is in conflict with the efficiency requirement of a low asymptotic variance under the parametric model." The optimal class of compromise statistics given by Lemma 5 of Hampel and its generalizations show that at least for M-estimators, increasing the bound on robustness (smaller gross error sensitivity) leads to a decrease in efficiency. After the introduction of Hampel's redescending psi function $\psi \equiv \psi_1$ of Eq. (1.21), the notion of *rejection point*, where the influence function vanishes, became prominent. Hampel et al. (1986) exploit all these criteria and more in their comprehensive book on this subject, including specific comparisons involving the calculus of variations allied with gross error sensitivity, rejection points, and many more optimality criteria.

The following subsections include theory detailed in Clarke (1986b).

3.2.1   Non-Smooth Analysis and Conditions A'

The loss functions L given in Eq. (3.2) and corresponding psi functions that emanate from it as illustrated by Figures 1.12 and 1.13 present problems in terms of the calculus that can lead to a proof of weak continuity of the estimating functional T and Fréchet differentiability with respect to the Kolmogorov metric. The class of functions $\{(\partial/\partial\tau)\Psi(x, \tau) | \tau \in D\}$ is not continuous, essentially because of the sharp corners in the component functions of Ψ. Condition A_1 of conditions **A** is not met. A similar observation can be made of the component functions of the location and scale redescenders derived from Hampel in Figures 1.10 and 1.11. Vector parameter proofs of weak continuity and Fréchet differentiability of M-functionals in Chapter 2 rely on the existence of continuous partial derivatives of the function Ψ, or in the case of location and scale estimation, continuously differentiable ψ_i, $i = 1, 2$. With component functions ψ_i where there exist points where there are "sharp corners" conditions **A** are not met. This is also a problem with general Ψ derived from generalizations of Hampel's Lemma 5, described in Hampel et al. (1986). Asymptotic expansions that make use of say the multivariate inverse function theorem and the multivariate mean value theorem are not then directly applicable given that $K_{F_n}(\tau)$ is not differentiable continuously in τ, even though it may be $K_{F_n}(\tau)$ remains continuous. At this point we could be said to be in harmony with the famous mathematician Hermite, who wrote in a letter to Stieltjes:

> Je me détournee avec effroi et horreur de cette plai lamentable des functions qui n'ont pas de dérivées.

See F.H. Clarke (1983b, p. 284). That book, however, deals with the calculus of non-smooth functions and offers insights and theorems, including the inverse function theorem and the multivariate mean value theorem for functions that are known to be Lipschitz but not necessarily smooth. These lend themselves to the theory of non-smooth analysis vis-à-vis the discussion in F. H. Clarke 1983b. For example, let $f : \mathcal{E}^r \rightarrow \mathcal{E}^r$ be a function, and let x be a point in $\mathcal{E}^r$. The function f is said to be Lipschitz near x if there exists a scalar C and a positive number ϵ such that the following holds:

$$|f(x'') - f(x')| \leq C|x'' - x'| \quad \text{for all } x'', x' \text{ in } x + \epsilon B$$

(Here B denotes the open unit ball in $\mathcal{E}^r$, so that $x + \epsilon B$ is the open ball of radius ϵ about x.)

Suppose θ is a point near which f is Lipschitz. Denote Ω_f to be the set of all points at which f fails to be differentiable. Rademacher's theorem states that a function which is Lipschitz on an open set of $\mathcal{E}^r$ is differentiable almost everywhere (a.e.) (in the sense of Lebesgue measure) on that subset. Hence, Ω_f is known to be a set of Lebesgue measure zero. Let $Jf(\tau)$ be the usual $r \times r$ matrix of partial derivatives whenever $\tau \notin \Omega_f$.

Definition 3.1: The generalized Jacobian of f at θ, denoted by $\partial f(\theta)$ is the convex hull of all $r \times r$ matrices Z obtained as a limit of a sequence of the form $Jf(\tau_i)$ where $\tau_i \rightarrow \theta$ and $\tau_i \notin \Omega_f$.

The generalized Jacobian $\partial f(\theta)$ is said to be of maximal rank provided every matrix in $\partial f(\theta)$ is of maximal rank (i.e. non-singular). The following proposition is proved on page 71 of Clarke (1983b). Propositions 3.1, 3.2, and 3.3 are with permission from F. H. Clarke.

Proposition 3.1: (F. R. Clarke, 1983) The generalized Jacobian $\partial f(\theta)$ is upper semi-continuous, which means, given $\epsilon > 0$ there exists a $\delta > 0$ such that for all $\tau \in U_\delta(\theta)$, the open ball of radius δ centered at θ

$$\partial f(\tau) \subset \partial f(\theta) + \epsilon B_{r \times r}.$$

Here $B_{r \times r}$ is the unit ball of matrices for which $B \in B_{r \times r}$ implies $||B|| \leq 1$.   □

It can be remarked that without loss of generality $||B||$ can be the least upper bound of $|By|$ where $|y| \leq 1$.

Conditions **A'** which follow are sufficient to establish for resultant functionals $T[\cdot]$, weak continuity with respect to the Prohorov distance metric, and Fréchet differentiability with respect to the Kolmogorov distance metric. We again take the path of Chapter 2 where we define an auxiliary functional $\rho(G, \tau) = |\tau - \theta|$, show existence of suitable functionals $T[\Psi, \rho, \cdot]$, and then for suitable robust selection functionals ρ_o these properties of weak continuity and Fréchet differentiability are retained according to Theorem 2.4. As in the previous discussion of Chapter 2 previously, we now state the conditions for a general ρ and T.

Conditions A'

A_0': $T[\Psi, \rho, F_\theta] = \theta$.

A_1': $\Psi(x, \tau)$ is an $r \times 1$ vector function on $R \times \Theta$ which is continuous and bounded on $R \times D$, where $D \subset \Theta$ is some nondegenerate compact interval containing θ in its interior, and R is some separable metrizeable space.

A_2': $\Psi(x, \tau)$ is locally Lipschitz in τ about θ in the sense that for some constant α

$$|\Psi(x, \tau) - \Psi(x, \theta)| < \alpha |\tau - \theta|$$

uniformly in $x \in R$ and for all τ in a neighborhood of θ.

A_3': Letting differentiation be with respect to the argument in parentheses

$$\partial K_{F_\theta}(\tau) \text{ is of maximal rank at } \tau = \theta.$$

A_4': Given $\delta > 0$ there exists an $\epsilon > 0$ such that for all $G \in n(\epsilon, F_\theta)$

$$\sup_{\tau \in D} |K_G(\tau) - K_{F_\theta}(\tau)| < \delta$$

and

$$\partial K_G(\tau) \subset \partial K_{F_\theta}(\tau) + \delta B_{r \times r} \text{ uniformly in } \tau \in D.$$

Remark 3.1: $A_0' \equiv A_0$

Remark 3.2: For a given function Ψ satisfying A_1', it follows from Lemma 2.2 and Theorem 2.1 that given $\delta > 0$ there exists an $\epsilon > 0$ such that whenever $G \in n(\epsilon, F_\theta)$

$$\sup_{\tau \in D} |K_G(\tau) - K_{F_\theta}(\tau)| < \delta.$$

This can be shown to be true for neighborhoods generated by metrics d_k, d_L, and d_p. This establishes the first part of condition A_4'.

Remark 3.3: If $K_{F_\theta}(\tau)$ is continuously differentiable in τ at θ, then $A_3' \equiv A_3$.

Conditions A_0'–A_3' can be considered as fairly straightforward, whereas condition A_4' is not so obvious. When $R = \mathcal{E}$, the real line, it can be shown to be a consequence of the following theorem.

Theorem 3.1: *(Clarke 1986b, Theorem 2.1) Let $\mathcal{A}$ be a class of real vector continuous functions defined on $\mathcal{E}$ with the following properties:*

1. *$\mathcal{A}$ is uniformly bounded, that is, there exists a constant H such that $|f(x)| \leq H < \infty$ for all $f \in \mathcal{A}$ and $x \in \mathcal{E}$.*
2. *$\mathcal{A}$ is equicontinuous.*

Let $F_\theta \in \mathcal{F}$ be given. Then for every $\delta > 0$ there is an $\epsilon > 0$ such that $d_k(F_\theta, G) \leq \epsilon$ implies

$$\sup_{f \in \mathcal{A}} \sup_{x \in \mathcal{E} \cup \{+\infty\}} \left| \int_{I_x} f dG - \int_{I_x} f dF_\theta \right| < \delta,$$

where integration is performed componentwise over intervals I_x which can be either open or closed of the form $(-\infty, x)$ or $(-\infty, x]$.

The proof of this theorem is given for observation spaces contained in the real line $\mathcal{E}$ in the appendix of Clarke (1986b). It is an open problem to consider generalizations to vector observation spaces.

Consider Ψ with continuous and bounded partial derivatives bar on a finite set of points $S^*(\tau)$. From F.H. Clarke (1983b, pp. 75–83) it follows that

$$\partial K_G(\tau) = \partial \int \Psi(y, \tau) dG(y) \subset \int \partial \Psi(y, \tau) dG(y)$$

from which the right-hand side can be expanded to a finite summation

$$\sum_{j=1}^{m} \int_{I_j} f_j(y,\tau)dG(y) + \sum_{x\in S^*(\tau)} \partial\Psi(x,\tau)G\{x\}.$$

Here $f_j \in \mathcal{A}$ and $\frac{\partial\Psi}{\partial\tau}(y,\tau) = f_j(y,\tau)$ on the connected interval I_j, for $j = 1, \ldots, m$. Since Ψ is Lipschitz in τ and $\partial\Psi(x,\tau)$ bounded, Theorem 3.1 implies condition A_4'.

3.2.2   Existence and Uniqueness for Solutions of Equations

We need the following propositions established on pp. 252–255 of Clarke (1983b) to avoid the need to assume continuous partial derivatives in the argument for the inverse function theorem.

Proposition 3.2: Suppose f satisfies properties described in Section 3.2.1 and

$$4\lambda_f \leq \inf_{\partial f(\theta)} ||M(\theta,f)||,$$

where the infimum is taken over all matrices $M(\theta,f) \in \partial f(\theta)$, and for some $\delta > 0$, $\tau \in U_\delta(\theta)$ implies

$$2\lambda_f \leq \inf_{\partial f(\theta)} ||M(\tau,f)||.$$

Then for arbitrary $\tau_1, \tau_2 \in \overline{U}_\delta(\theta)$, the closure of the ball $U_\delta(\theta)$

$$|f(\tau_1)-f(\tau_2)| \geq 2\lambda_f|\tau_1-\tau_2|.$$

$\square$

Proposition 3.3: Under the conditions of Proposition 3.2 $f(U_\delta(\theta))$ contains $U_{\lambda_f\delta}(f(\theta))$. $\square$

Remark 3.4: For $v \in U_{\lambda_f\delta}(f(\theta))$ we can define $f^{-1}(v)$ to be the unique $\tau \in U_{\lambda_f\delta}(\theta)$ such that $f(\tau) = v$ and Proposition 3.2 implies f^{-1} is Lipschitz with Lipschitz constant $1/(2\lambda_f)$.

Lemma 3.1: *Let conditions A' hold for some Ψ, ρ. Then there exists a $\delta_1 > 0$ and an $\epsilon_1 > 0$ such that for all $G \in n(\epsilon_1, F_\theta)$ any matrix*

$$M(\tau,G) \in \partial K_G(\tau)$$

will satisfy $||M(\tau,G)|| > 2\lambda$, where λ is defined to be a value for which $M(\theta,F_\theta) \in \partial K_{F_\theta}(\theta)$ implies $||M(\theta,F_\theta)|| > 4\lambda$.

Remark 3.5: If $K_{F_\theta}(\tau)$ is continuously differentiable in τ then the choice of $\lambda = 1/(4\|M(\theta, F_\theta)^{-1}\|)$ satisfies the criterion of Lemma 3.1.

Proof. (of Lemma 3.1) Since ∂K_{F_θ} is upper semi-continuous, choose by Proposition 3.1 $\delta_1 > 0$ such that $\partial K_{F_\theta}(\tau) \subset \partial K_{F_\theta}(\theta) + \lambda B_{r\times r}$ whenever $\tau \in U_{\delta_1}(\theta)$. By condition A_4' there exists an $\epsilon_1 > 0$ such that $G \in n(\epsilon_1, F_\theta)$ implies

$$\partial K_G(\tau) \subset \partial K_{F_\theta}(\tau) + \lambda B_{r\times r} \text{ uniformly in } \tau \in D.$$

Hence, given $M(\tau, G) \in \partial K_G(\tau)$ for $\tau \in U_{\delta_1}(\theta)$, there exists $M(\theta, F_\theta) \in \partial K_{F_\theta}(\theta)$ such that

$$\|M(\tau, G) - M(\theta, F_\theta)\| < 2\lambda,$$

whence by Proposition 3.2 $\|M(\tau, G)\| > 2\lambda$. It is now possible to state and prove the existence and uniqueness argument analogous to Theorem 2.3 but using conditions **A'**. This result also proves existence of a weakly continuous root for either the Lévy or Prohorov neighborhoods. As usual the following selection functional is only used as an auxiliary device.

Theorem 3.2: *(Clarke, 1986b, Theorem 3.1) Let $\rho(G, \tau) = |\tau - \theta|$ and suppose conditions A' hold. Then given $\kappa > 0$ there exists an $\epsilon > 0$ such that $G \in n(\epsilon, F_\theta)$ implies $T[\Psi, \rho, G]$ exists and is an element of $U_\kappa(\theta)$. Further for this ϵ there is a $\kappa^* > 0$ such that*

$$S(\Psi, G) \cap U_{\kappa^*}(\theta) = T[\Psi, \rho, G],$$

and $\partial K_G(\tau)$ is of maximal rank for $\tau \in U_{\kappa^*}(\theta)$. For any null sequence of positive numbers $\{\epsilon_k\}$, let $\{G_k\}$ be an arbitrary sequence for which $G_k \in n(\epsilon_k, F_\theta)$. Then

$$\lim_{k\to\infty} T[\Psi, \rho, G_k] = T[\Psi, \rho, F_\theta] = \theta.$$

Proof. Since $\partial K_{F_\theta}(\tau)$ is upper semi-continuous in τ, choose $0 < \kappa^* < \min(\delta_1, \kappa)$ such that $\tau \in U_{\kappa^*}(\theta)$ implies

$$\inf_{\partial K_G(\tau)} \|M(\tau, G)\| > 2\lambda \text{ for all } G \in n(\epsilon_1, F_\theta)$$

where the infimum is taken over all matrices $M(\tau, G) \in \partial K_G(\tau)$. Here δ_1, ϵ_1, and λ are defined in Lemma 3.1. Hence,

$$4\lambda(G) = \inf_{\partial K_G(\theta)} \|M(\theta, G)\| > 2\lambda.$$

Choose $0 < \epsilon^* \leq \epsilon_1$ so that the following relations hold.

$$\partial K_G(\tau) \subset \partial K_{F_\theta}(\tau) + (\lambda/4)B_{r \times r} \text{ by } A_4'$$

$$\subset \partial K_{F_\theta}(\theta) + (\lambda/2)B_{r \times r} \text{ by Proposition 3.1}$$

$$\subset \partial K_G(\theta) + \lambda B_{r \times r} \text{ by } A_4'.$$

Then for every $M(\tau, G) \in \partial K_G(\tau)$ there exists an $M(\theta, G) \in \partial K_G(\theta)$ such that

$$\|M(\tau, G) - M(\theta, G)\| < \lambda < 2\lambda(G),$$

whenever $G \in n(\epsilon^*, F_\theta)$ and uniformly in $\tau \in U_{\kappa^*}(\theta)$. By Proposition 3.2 $K_G(\cdot)$ is a one-to-one function from $U_{\kappa^*}(\theta)$ onto $K_G(U_{\kappa^*}(\theta))$ and by Proposition 3.3 the image set contains the open ball of radius $\lambda \kappa^*/2$ about $K_G(\theta)$. The argument for uniqueness is now the same as in the proof of Theorem 2.3.

3.2.3   Results for M-estimators with Non-Smooth Ψ

It is assumed in this section that $K_{F_\theta}(\tau)$ has at least a continuous derivative $\frac{\partial}{\partial \tau} K_{F_\theta}(\tau)$ at $\tau = 0$, which is denoted by $M(\theta)$. This is common with absolutely continuous parametric families. With this restriction Fréchet differentiability follows.

Theorem 3.3: *(Clarke, 1986b, Theorem 4.1) Let $\rho(G, \tau) = |\tau - \theta|$ and assume conditions A' hold with respect to the functional $T[\Psi, \rho, \cdot]$ and neighborhoods generated by the metrics d_* on $\mathcal{G}$. Suppose for all $G \in \mathcal{G}$*

$$\left| \int_R \Psi(x, \theta) \mathrm{d}(G - F_\theta)(x) \right| = O(d_*(G, F_\theta)) \tag{3.4}$$

as $d_*(G, F_\theta) \to 0$. Then $T[\Psi, \rho, \cdot]$ is Fréchet differentiable at F_θ with respect to $(\mathcal{G}, d_*)$ and has derivative

$$T'_{F_\theta}(G - F_\theta) = -M(\theta)^{-1} \int_R \Psi(x, \theta) \mathrm{d}(G - F_\theta)(x).$$

To prove the theorem, it is necessary to introduce the following generalization of the mean value result described as Proposition 2.6.5 in F.H. Clarke 1983b.

Proposition 3.4: (Mean value theorem) Let f be Lipschitz on an open convex set U in $\mathcal{E}^r$ and let τ_1 and τ_2 be points in U. Then one has

$$f(\tau_1) - f(\tau_2) \in \mathrm{co}\partial f([\tau_1, \tau_2])(\tau_2 - \tau_1).$$

(The right-hand side above denotes the convex hull of all points of the form $Z(\tau_2 - \tau_1)$ where $Z \in \partial f(u)$ for some point u in $[\tau_1, \tau_2]$. Since $[\mathrm{co}\partial f([\tau_1, \tau_2])](\tau_2 - \tau_1) = \mathrm{co}[\partial f([\tau_1, \tau_2])(\tau_2 - \tau_1)]$, there is no ambiguity.)

Proof. (of Theorem 3.3) Abbreviate $T[\Psi, \rho, \cdot] = T[\cdot]$ and let κ^*, ϵ be given by Theorem 2.3. Let $\{\epsilon_k\}$ be so that $\epsilon_k \downarrow 0^+$ as $k \to \infty$ and let $\{G_k\}$ be any sequence such that $G_k \in n(\epsilon_k, F_\theta)$. By Theorem 3.2, $T[G_k]$ exists and is unique in $U_{\kappa^*}(\theta)$ for $k > k_o$ where $\epsilon_{k_o} \leq \epsilon$. By A'_4 see that for arbitrary $\delta > 0$

$$\partial K_{G_k}(\tau) \subset \partial K_{F_\theta}(\tau) + \delta B_{r \times r} \quad \text{uniformly in } \tau \in D \tag{3.5}$$

for sufficiently large k. Consider the two-term expansion,

$$0 = K_{G_k}(T[G_k]) = K_{G_k}(\theta) + M(\tilde{\tau}_k, G_k)(T[G_k] - \theta), \tag{3.6}$$

where $|\tilde{\tau}_k - \theta| < |T[G_k] - \theta|$, which tends to zero as $k \to \infty$ by Theorem 3.2, and $\tilde{\tau}_k$ is evaluated at different points for each component function expansion obtained as a consequence of Proposition 3.4 (i.e. $\tilde{\tau}_k$ takes different values in each row of matrix M). See from Eqs. (3.6), (3.4), and Lemma 3.1 that

$$|T[G_k] - \theta| = O(K_{G_k}(\theta)) = O(\epsilon_k).$$

Also,

$$T[G_k] - \theta = -M(\theta)^{-1}K_{G_k}(\theta) - M(\theta)^{-1}\{M(\tilde{\tau}_k, G_k) - M(\theta)\}(T[G_k] - \theta).$$

By upper semi-continuity of $K_G(\tau)$ in τ and Eq. (3.5)

$$||M(\tilde{\tau}_k, G_k) - M(\theta)|| = o(1).$$

So

$$|T[G_k] - \theta - T'_{F_\theta}(G_k - F_\theta)| = o(1)O(d_*(G_k, F_\theta)) = o(\epsilon_k).$$

$\square$

Example 3.1: Consider Huber's Proposal 2 estimators which are solutions of equations (1.20) where the function $\Psi = (\psi_1, \psi_2)^T$ is given by Eq. (1.24). Setting $\theta = (\theta_1, \theta_2)^T$ and denoting $F_\theta(x) = \Phi\left(\frac{x - \theta_1}{\theta_2}\right)$ to be the normal distribution with location parameter θ_1 and scale parameter θ_2, then the vector function $K_{F_\theta}(\tau)$ is continuously differentiable in τ. Evaluating the derivative at θ, the resulting Jacobian is

$$M(\theta) = -\frac{1}{\theta_2}\begin{pmatrix} \int \psi'_1(y)\mathrm{d}\Phi(y) & 0 \\ 0 & \int y\psi'_2(y)\mathrm{d}\Phi(y) \end{pmatrix}. \tag{3.7}$$

Condition A_0' follows since $E_\Phi[\Psi] = 0$. A_1', A_2' hold by inspection, and A_3' holds since $M(\theta)$ is non-singular. Remark 3.2 suffices for the first part of A_4'. To apply Theorem 3.1, consider the function

$$
f(x, \tau) = I_{(\tau_1 - k\tau_2, \tau_1 + k\tau_2)} \begin{pmatrix} -\dfrac{1}{\tau_2} & \dfrac{-(x - \tau_1)}{\tau_2^2} \\ \dfrac{-(x - \tau_1)}{\tau_2^2} & \dfrac{-(x - \tau_1)^2}{\tau_2^3} \end{pmatrix}
$$
$$
+ \frac{1}{\tau_2}(I_{(-\infty, \tau_1 - k\tau_2]}(x) + I_{[\tau_1 + k\tau_2, \infty)}(x)) \begin{pmatrix} -1 & -k \\ -k & -k^2 \end{pmatrix}.
$$

It is clear that $\mathcal{A} = \{f(\cdot, \tau) : \tau \in D\}$ forms a bounded equicontinuous class of functions on $\mathcal{E}$. Also

$$
\partial K_G(\tau) = \int_{(\tau_1 - k\tau_2, \tau_1 + k\tau_2)} f(x, \tau) \, dG(x)
$$
$$
+ \partial\Psi(\tau_1 - k\tau_2, \tau) \, G\{\tau_1 - k\tau_2\} + \partial\Psi(\tau_1 + k\tau_2, \tau) \, G\{\tau_1 + k\tau_2\},
$$

where differentiation of Ψ is with respect to the second argument, while

$$
\partial K_{F_\theta}(\tau) - \int_{(\tau_1 - k\tau_2, \tau_1 + k\tau_2)} f(x, \tau) \, dF_\theta(x).
$$

A_4' follows by Theorem 3.1 and because $\partial\Psi(\tau_1 - k\tau_2, \tau)$ and $\partial\Psi(\tau_1 + k\tau_2, \tau)$ are bounded. Assumption (3.4) holds for the Kolmogorov distance metric through integration by parts and noting that Ψ is a function of total bounded variation. Thus, by Theorems 3.2 and 3.3 there exists a root that is Fréchet differentiable with respect to $(\mathcal{G}, d_k)$ at the normal distribution F_θ. It is also Fréchet differentiable at the distribution $F_o(\frac{x - \theta_1}{\theta_2})$ for which the density function of F_o is

$$
f_o(x) = \begin{cases} \dfrac{(1-\epsilon)}{\sqrt{2\pi}} e^{-\frac{x^2}{2}} & \text{for } |x| \le k \\[2ex] \dfrac{(1-\epsilon)}{\sqrt{2\pi}} e^{\frac{k^2}{2} - k|x|} & \text{for } |x| > k \end{cases}
$$

with k and ϵ connected through Eq. (3.3). Then the M-estimator coincides with the maximum likelihood estimator and provides another example of a robust and asymptotically efficient estimator based on Huber's least favorable distribution.□

3.3   REGRESSION M-ESTIMATORS

Almost a decade after the introduction of M-estimators of location, Huber (1973) introduced M-estimators into regression. Given a model

$$Y_i = \beta_o + \beta_1 x_{1i} + \cdots + \beta_m x_{mi} + \epsilon_i; i = 1, \ldots, n, \tag{3.8}$$

where $\epsilon_i \sim N(0, \sigma^2)$ and an estimated parameter $\hat{\beta} = (\hat{\beta}_o, \hat{\beta}_1, \ldots, \hat{\beta}_m)^T$ which is $(m+1) \times 1$ or $p \times 1$ say, the ith estimated residual can be written as

$$\epsilon_i(\hat{\beta}) = Y_i - \hat{\beta}_o - \hat{\beta}_1 x_{1i} - \cdots - \hat{\beta}_m x_{mi}. \tag{3.9}$$

The proposal of Huber (1973) following the lead of Gauss and Legendre was to choose $\hat{\beta}$ to solve

$$\sum_{i=1}^{n} L(\epsilon_i(\beta)) = \min! \tag{3.10}$$

Naturally, the choice for the distribution of ϵ_i as $N(0, \sigma^2)$ leads to classical estimation and inference for $\hat{\beta}$ through the theory of least squares estimation with $L(x) = x^2$. The book of Clarke (2008) is devoted to this, for example. However, Huber's proposal was to consider estimators of regression and scale parameters through solving the minimizing equations

$$\sum_{i=1}^{n} \psi_1 \left(\frac{\epsilon_i}{\hat{\sigma}} \right) x_i = 0 \tag{3.11}$$

$$\sum_{i=1}^{n} \psi_2 \left(\frac{\epsilon_i}{\hat{\sigma}} \right) = 0, \tag{3.12}$$

where $x_i^T = [1, x_{1i}, \ldots, x_{mi}]$. His suggestion was to choose $\Psi = (\psi_1, \psi_2)$ according to Huber's Proposal 2 given in formula (1.24). Such estimators are robust to the choice of actual distribution of ϵ_i, in particular allowing for example that the ϵ_i may not in fact be $N(0, \sigma^2)$ random variables but may be sampled through some heavier tailed but symmetric distribution, for example.

However, Huber's proposed M-estimator, being a solution of Eqs. (3.11) and (3.12), does not cope with grossly aberrant values of x_i, which have a large influence on the fit. The vulnerability of these M-estimators to leverage points eventually led to generalized M-estimators (GM-estimators), a version due to Schweppe (see Hill (1977)) uses

$$\sum_{i=1}^{n} w(x_i)\psi_1 \left(\frac{\epsilon_i}{(w(x_i)\hat{\sigma})} \right) x_i = 0. \tag{3.13}$$

Leverage points are explained in chapter 7 of Clarke (2008), for example. We consider the functional approach as elegantly relayed in Hampel et al. (1986, pp. 315–317) to this estimation problem. The pair (x^T, y) can be thought of as coming from their own distribution, F say. We study the influence function of the more general M-functional at the distribution $F_\beta(x, y)$ with density (ignoring the scale)

$$f_\beta(x, y) = \phi(y - x^T\beta)k^*(x).$$

The functional approach to the more general problem is to consider M-estimators T_n for linear models defined as solutions of the vector equation (3.14). Quoting from Hampel et al. (1986, section 6.3, p. 315) with permission beginning with Definition 1, "An M-estimator T_n for linear models is defined implicitly by the (vector) equation

$$\sum_{i=1}^{n} \eta \left(x_i, \frac{y_i - x_i^T T_n}{\sigma} \right) x_i = 0 \tag{3.14}$$

(See Maronna and Yohai (1981)), where the function $\eta : \mathcal{E}^p \times \mathcal{E} \to \mathcal{E}$ satisfies the following conditions:

ETA'_1
    (i) $\eta(x, \cdot)$ is continuous on $\mathcal{E}\backslash C(x; \eta)$ for all $x \in \mathcal{E}^p$ where $C(x; \eta)$ is a finite set. In each point of $C(x; \eta)$, $\eta(x, \cdot)$ has finite left and right limits.
    (ii) $\eta(x, \cdot)$ is odd and $\eta(x, r) \geq 0$ for all $x \in \mathcal{E}^p$, $r \in \mathcal{E}^+$.
ETA'_2 For all x, the set $D(x; \eta)$ of points in which $\eta(x, \cdot)$ is continuous but in which $\eta'(x, r) = (\partial/\partial r)\eta(x, r)$ is not defined or not continuous is finite.
ETA'_3
    (i) $M = E\eta'(x, r)xx^T = \int \eta'(x, r)xx^T d\Phi(r)dK^*(x)$ exists and is nonsingular.
    (ii) $Q = E\eta^2(x, r)xx^T = \int \eta^2(x, r)xx^T d\Phi(r)dK^*(x)$ exists and is nonsingular."

Hampel et al. (1986, section 6.3, pp. 315–317) further write and we quote with permission: "All known proposals of η may be written in the form

$$\eta(x, r) = w(x) \cdot \psi(r \cdot v(x))$$

for appropriate functions $\psi : \mathcal{E} \to \mathcal{E}$ and $w : \mathcal{E}^p \to \mathcal{E}^+$, $v : \mathcal{E}^p \to \mathcal{E}^+$ (weight functions). Huber (1973) uses $w(x) = 1$, $v(x) = 1$." And then "Let now $\sigma = 1$ for simplicity. The functional $T[F]$ corresponding to the M-estimator of Eq. (3.14) is the solution of

$$\eta(x, y - x^T \cdot T[F])x \mathrm{d}F(x, y) = 0 \qquad (3.15)$$

Define

$$M(\eta, F) = \int \eta'(x, y - x^T \cdot T[F])x \cdot x^T \mathrm{d}F(x, y) \qquad (3.16)$$

Then the influence function of T at a distribution F (on $\mathcal{E}^p \times \mathcal{E}$) is given by

$$\mathrm{IF}(x, y; T, F) = \eta(x, y - x^T \cdot T[F]) \cdot M^{-1}(\eta, F) \cdot x \qquad (3.17)$$

Maronna and Yohai (1981) show, under certain conditions, that these estimators are consistent and asymptotically normal with asymptotic covariance matrix

$$\Sigma(T, F) = \int \mathrm{IF}(x, y; T, F) \cdot \mathrm{IF}^T(x, y; T, F) \, \mathrm{d}F(x, y)$$

$$= M^{-1}(\eta, F) \cdot Q(\eta, F) \cdot M^{-1}(\eta, F), \qquad (3.18)$$

where

$$Q(\eta, F) = \int \eta^2(x, y - x^T.T[F])xx^T \mathrm{d}F(x, y)$$

(cf. Yohai and Maronna (1979)). At the model distribution $F = F_\beta$ we obtain

$$\mathrm{IF}(x, y; T, F_\beta) = \eta(x, y - x^T\beta) \cdot M^{-1} \cdot x,$$

$$\Sigma(T, F_\beta) = M^{-1}QM^{-1},$$

where

$$M = M(\eta, \Phi) = E\{\eta'(x, \epsilon)xx^T\}$$

$$Q = Q(\eta, \Phi) = E\{\eta^2(x, \epsilon)xx^T\}.$$

Note that M and Q, and thus $\Sigma(T, F_\beta)$ do not depend on β." Hampel et al. (1986) go on to describe the sensitivity in various guises and the influence on the fitted value or self influence as in Hampel (1978).

Müller (1997) proves Fréchet differentiability, with respect to the Kolmogorov distance metric, of Huber's M-functional for linear models under some technical conditions that she verifies using Example 3.1 initially given in Clarke (1986b), and illustrates her results for planned experiments.

Unaware of Müller's work, Christmann and Van Messem (2008) in a separate development introduce what is termed a Bouligand derivative for use in establishing robustness of support vector machines (SVMs) for regression that employ among others Huber's convex loss function of Eq. (3.2). An, SVM is a tool used in machine learning. Since computers have difficulty with complex tasks (such as converting hand writing into plain text), they use huge amounts of data to perform reasonably well. The driving force behind SVMs has been the need to cut down processing time, and use the data as efficiently as possible. What is termed a sparseness property implies that only a fraction of the data are used to compute the final estimate. When this is employed, the data points that contribute directly to the answer are called the support vectors. SVMs use loss functions instead of ψ-functions. Fréchet differentiability implies both Bouligand and Gâteaux differentiability. A Bouligand derivative need not be linear. However, in the instance of Huber's loss function, which leads to an M-functional that is Fréchet differentiable according to the results of Müller (1997), it would appear there is no gain in extending the notions of differentiability here, albeit the mathematics is interesting.

3.4 STOCHASTIC FRÉCHET EXPANSIONS AND FURTHER CONSIDERATIONS

A principal aim in establishing expansions such as the Fréchet expansion (2.42) or the more general expansion (2.42) is, at least for independent identically distributed random variables, to make a substitution of the empirical distribution F_n for F and the parametric distribution F_θ for G and subsequently argue the remainder is $o_p(1/\sqrt{n})$, whenever F_n is generated by F_θ. That is, one arrives at the stochastic form of the Fréchet expansion which is

$$T[F_n] - T[F_\theta] - \int \text{IF}(x, F_\theta, T) \text{d}(F_n - F_\theta)(x) + o_p\left(\frac{1}{\sqrt{n}}\right). \qquad (3.19)$$

As argued in Huber (1981, p. 39), and Huber and Ronchetti (2009, p. 140), for any estimating functional T that is Fréchet differentiable at F_θ and Fisher consistent in the sense that $T[F_\theta] = \theta$ Eq. (3.19) follows. See also expansion (2.32) for even multivariate data. The expansion (3.19) was first exploited for M-estimators by Boos and Serfling (1980). See also Serfling (1980). It does not require Fréchet differentiability and is usually arrived at by the δ-method. For maximum likelihood estimation, the first rigorous arguments for consistency, $(\hat{\theta} \to_p \theta)$, and asymptotic normality, $\sqrt{n}(\hat{\theta} - \theta) \to_d N(0, \mathcal{I}(\theta)^{-1})$, were detailed under the celebrated Cramér Conditions of Cramér (1946). Generalizations of such conditions appear in many places, the most recent being for M-estimators, both univariate and multivariate, in Bickel and Doksum (2007). They also detail asymptotic results for M-estimators of regression parameters. It is emphasized that expansions such as Eq. (3.19) hold under very weak conditions, for instance, the IF(x, F_θ, T) need not be bounded in x, nor continuous in x, and hence Fréchet differentiability need not hold, and indeed the estimating functional need not be robust. Achieving expansion (3.19) for a wide variety of estimating functionals including M-estimators such as Huber's skipped mean for location defined by $\psi_k(x) = xI_{[-k,k]}(x)$, and the median $\psi_M(x) = \text{sign}(x)$, and L-estimators, which are linear combinations of order statistics, and R-estimators which are statistical estimators based on ranks, under a variety of conditions, is a major focus of the book of Jurečková et al. (2012). See Clarke (2014) for an overview. It is not intended to emulate their work here albeit it is important.

3.5   LOCALLY UNIFORM FRÉCHET EXPANSION

It can be noted, however, that expansions in neighborhoods of F_θ are facilitated by arguments of Fréchet differentiability. In what might be a defining work in this regard involving $\sqrt{n}$-shrinking neighborhoods, Bednarski et al. (1991) show that estimators which have locally uniform expansions are asymptotically equivalent to M-estimators and that the M-functionals corresponding to these M-estimators are seen to be locally uniformly Fréchet differentiable. One considers infinitesimal models that are given in Jaeckel (1971a), Beran (1977), Huber-Carol (1970), and Rieder (1978), a justification for which can be found in Bickel (1981) and in Bednarski (1993a). Let $\mathcal{F}_{\theta,n}$ denote the shrinking neighborhood about F_θ, and let T_n be an estimator of θ. Generalizing Eq. (3.19), Bickel (1981), Rieder (1978), and Bednarski (1985) have that, for some function Ψ,

$$\sqrt{n}(T_n - \theta) = \frac{1}{\sqrt{n}} \sum_{i=1}^{n} \Psi(X_i, \theta) + o_{F_\theta^{\otimes n}}(1). \qquad (3.20)$$

Here $X_1, \dots, X_n$ is a sample from F_θ and estimators satisfying Eq. (3.20) are called regular. Bickel and Rieder consider analysis of regular estimators under contiguous departures from the basic model, whereby expansions hold automatically for product measures from $\mathcal{F}_{\theta,n}^{\otimes n}$. Bednarski et al. (1991) consider an important subclass of regular estimators, where expansions are locally uniform so that

$$\sqrt{n}(T_n - \tau_n) = \frac{1}{\sqrt{n}} \sum_{i=1}^{n} \Psi(X_i, \tau_n) + o_{Q_n}(1) \qquad (3.21)$$

for every sequence $Q_n \in \mathcal{F}_{\tau_n,n}^{\otimes n}$, where τ_n converges to θ so that $\sqrt{n}(\tau_n - \theta)$ stays bounded. T_n satisfying Eq. (3.21) are said to have strong expansions at F_θ with respect to the neighborhoods. Bednarski et al. (1991) make use of a fixed point theorem to show that if an estimator has a strong expansion (at F_θ with respect to neighborhoods generated by a metric distance d), with a function Ψ, then under additional mild assumptions, the M-functional related to Ψ is locally uniformly Fréchet differentiable, meaning that

$$|T[G] - \tau - M^{-1} \int \Psi(x,\tau)dG(x)| = o(d(F_\theta, F_\tau) + d(F_\tau, G)) \qquad (3.22)$$

for some matrix M. For an M-estimator Eq. (3.21) is implied by Eq. (3.22). Conditions for locally uniform Fréchet differentiability of M-estimators are related in Bednarski et al. (1991) and these are shown to be implied by the assumptions of Clarke (1983a) and Clarke (1986b) discussed above, whence the "usual" M-estimators with bounded and continuous Ψ functions have strong expansions.

Bednarski et al. (1991) illustrate that some well-known estimators which are robust in the sense that they are resistant to large outliers need not at the same time have a strong expansion. In particular, as has already been noted, the median is not Fréchet differentiable. Denoting the median to be $F_n^{-1}(\frac{1}{2})$ and noting the influence function (compare with Eq. (2.43)) for the median at the normal distribution Φ is given by a discontinuous function

$$\psi(x) = \left\{2\phi\left(\Phi^{-1}\left(\tfrac{1}{2}\right)\right)\right\}^{-1} \text{sign}\left(x - \Phi^{-1}\left(\tfrac{1}{2}\right)\right),$$

the expansion

$$\sqrt{n}F_n^{-1}\left(\tfrac{1}{2}\right) = \frac{1}{\sqrt{n}}\sum_{i=1}^n \psi(X_i) + o_{\Phi\otimes n}(1)$$

is true at the normal distribution and the asymptotic normal distribution of $N(0, 1/\{4\phi(0)\}^2)$ is achieved. On the other hand, the strong expansion does not hold in view of the following lemma.

Lemma 3.2: *(Bednarski et al., 1991, Lemma 3.1) Let $X_1, \ldots, X_n$ be independent identically distributed random variables from*

$$G_n = \left(1 - \frac{\epsilon}{\sqrt{n}}\right)\Phi + \frac{\epsilon}{\sqrt{n}}\delta_o,$$

where δ_o places atomic mass one at the origin and $0 < \epsilon < 1$. Then asymptotically $\sqrt{n}F_n^{-1}\left(\tfrac{1}{2}\right)$ has a distribution that attributes positive mass at the origin, while $\left(1/\sqrt{n}\right)\sum_{i=1}^n \psi(X_i)$ tends in distribution to a normal variable.

Expansions (3.21) and (3.22) are valid under violations of the model (mild in terms of supremum norm or Kolmogorov distance), which may represent gross errors or asymmetry of the underlying distribution, for example. Hence, reliable estimates of the variance of such estimators can easily be constructed. More explicitly, suppose that the functional T is Fréchet differentiable at a distribution F defined on $\mathcal{E}^k$, with respect to the pair $(\mathcal{G}, d)$ with a functional derivative T'_F that can be expressed as an integral in the form $T'_F(G - F) = \int \psi_F(x) d(G - F)(x)$. Then

$$T[G] - T[F] = \int \psi_F dG(x) + o(d(G, F)),$$

as $d(G, F) \to 0$. Hence, it is easily seen, for G_n satisfying $d(G_n, F) \le c/\sqrt{n}$ and F_n the empirical distribution function based on the sample drawn from G_n, that $\sqrt{n}\{T[F_n] - T[G_n]\}$ is asymptotically normal $N(0, \sigma^2)$, where $\sigma^2 = E_F \psi^2$ and that $\int \psi^2 dF_n$ converges to σ^2 (uniformly for distributions G_n in $\sqrt{n}$ shrinking neighborhoods). This is very important because it enables the construction of more reliable tests and confidence intervals of parameters in the model. Also importantly, since T'_F has to be continuous with respect to the metric d, the function ψ_F has to be bounded. Further work in related directions is given in Bednarski and Clarke (1998).

3.6   CONCLUDING REMARKS

The idea of the "optimality" of maximum likelihood estimators in terms of their "efficiency" promulgated initially in Fisher (1925) was seriously brought into question by an example due to J.L. Hodges, Jr., which was reported in Le Cam (1953). It concerns the $N(\theta, 1)$ model, where the distribution $F_\theta(x) = \Phi(x - \theta)$. Suppose $X_1, \dots, X_n$ are independent random variables from this model. It is well known that $\overline{X}_n$ is the maximum likelihood estimate of the mean with variance $\sigma_n^2 = 1/n$. It is unbiased and attains the Cramér–Rao lower bound for all n. The expected information for this model is $\mathcal{I}(F_\theta) = 1$ for all $\theta \in \Theta$. As $n \to \infty$,

$$\sqrt{n}(\overline{X}_n - \theta_0) \to_d N(0, 1). \tag{3.23}$$

Defining the estimator $T_n(X1, \dots, X_n)$ via

$$T_n = \begin{cases} \overline{X}_n & \text{when } |\overline{X}_n| \ge n^{-1/4} \\ \alpha \overline{X}_n & \text{when } |\overline{X}_n| < n^{-1/4}, \end{cases} \tag{3.24}$$

it is an exercise to show that

$$\sqrt{n}(T_n - \theta_0) \to_d \begin{cases} N(0, 1) & \text{when } \theta_0 \ne 0, \\ N(0, \alpha^2) & \text{when } \theta_0 = 0. \end{cases} \tag{3.25}$$

Since α is arbitrary, the asymptotic variance can be less than that of the maximum likelihood estimator when $\theta_0 = 0$. Based on the criterion of asymptotic variance alone and we now have an example of an estimator for which

$$\Sigma(T, F_\theta) \leq I(F_\theta)^{-1}$$

with strict inequality holding at $\theta = 0$. Both estimators $\overline{X}_n$ and T_n are asymptotically unbiased and asymptotically normal.

Small (2010, p. 170) writes: "The superefficient estimator is not a serious contenders as good estimators of θ ... Estimators such as T_n acquire a point of superefficiency by 'borrowing' efficiency from neighbouring values. As a result, superefficiency at a point such as $\theta = 0$ above leads to poor behaviour in a neighbourhood of that point." He goes on to discuss the philosophical implications of superefficient estimators.

In a dinner conversation with Professor Kallianpur in 1983 when the author was visiting UNC at Chapel Hill, it was explained that Fisher was visiting the Calcutta Statistical Institute soon after these questions of efficiency were raised in the Berkeley School, and Kallianpur and Rao decided to write their article introducing Fréchet differentiability to the statistical literature as a way of supporting Fisher's notions of efficiency of the maximum likelihood estimator. Of course, as we have seen, maximum likelihood estimators are not Fréchet differentiable for every parametric family, and it is only when they are, that the resulting estimator is robust and having a bounded and continuous influence function. Thus we have conditions **A**, and for those M-estimators with influence functions that are continuous but have "sharp" corners, conditions **A '**. Naturally, conditions **A '** are weaker than conditions **A**, but conditions **A** are easier to check when they are satisfied. In addition, Hampel's Theorem 2 indicating robustness of an estimator at each F in a small neighborhood of F_θ is applicable under conditions **A** according to Clarke (2000a), whereas robustness only at F_θ is guaranteed by conditions **A '**. Apparently, contamination near the sharp corners can have an effect on estimators, and Vecchia et al. (2012) advocate the choice of smooth influence functions for similar reasons.

Since σ is arbitrary, the asymptotic variance can be less than that of the maximum likelihood estimator when $\theta_0 = 0$. Hasa'd on the criterion of asymptotic variance alone and we now have an example of an estimator for which

4

MULTIPLE ROOTS

4.1 INTRODUCTION TO MULTIPLE ROOTS

The idea that there can be more than one root to one's estimating equations is perhaps perturbing to many. Huber's insight to introduce convex loss functions seemed to obviate the need to worry about this problem, and it can be said that the dominance of Huber's minimax solution was perhaps not swayed as much by the introduction of the Hampel three-part redescender due to the fact that the choice of the latter's psi-function can yield multiple roots to ones estimating equations. This is despite the obvious philosophical advantages of redescenders. Criticisms can be found in Rey (1977) and Birch (1980), for example. But the question is much deeper than just in the estimation of location or maybe even the study of regression. It in fact pervades the whole of maximum likelihood estimation and indeed the study of estimation methodology more generally. In this chapter we continue the study of asymptotic convergence, and especially in the case when asymptotically there can be multiple roots to one's estimation equations. Since the examples include both non-robust and robust estimators, classical and modern scenarios, we incorporate the discussion under the guise of conditions **A**, assuming continuously differentiable psi functions, and we introduce in the following section two theorems from Clarke et al. (1993) which elucidate the asymptotics.

Robustness Theory and Application, First Edition. Brenton R. Clarke.
© 2018 John Wiley & Sons, Inc. Published 2018 by John Wiley & Sons, Inc.
Companion website: www.wiley.com/go/clarke/robustnesstheoryandapplication

4.2   ASYMPTOTICS FOR MULTIPLE ROOTS

Assume the observation space $R \subset \mathcal{E}^k$ so that the estimating equations are of the general form

$$K_{F_n}(\tau) = \frac{1}{n} \sum_{j=1}^{n} \Psi(X_j, \tau) = 0, \qquad (4.1)$$

where F_n is the empirical distribution function attributing mass $1/n$ to each of the points $X_1, \dots, X_n$, and the parametric family of distributions is represented by $\mathcal{F} = \{F_\tau | \tau \in \Theta \subset \mathcal{E}^r\}$. For a generating distribution F_{θ_0}, the corresponding asymptotic equations are

$$K_{F_{\theta_0}}(\tau) = \int_R \Psi(x, \tau) \mathrm{d} F_{\theta_0}(x) = 0. \qquad (4.2)$$

More generally, to reiterate Eq. (2.2), we have that the estimating functional $T[\Psi, G]$ is a functional root of the equations

$$K_G(\tau) = \int_R \Psi(x, \tau) \mathrm{d} G(x) = 0, \qquad (4.3)$$

where $G \in \mathcal{G}$. If no solution exists, we can define $T[\Psi, G]$ as $+\infty$. In the event that several solutions of the equations (4.3) exist, we consider all possible solutions

$$S(\Psi, G) = \{\tau | K_G(\tau) = 0; \ \tau \in \Theta\}. \qquad (4.4)$$

Theoretically, the solution set S is well defined, and again to reiterate Eq. (2.3) we define the estimating functional uniquely via a selection functional ρ_o through the implicit equation

$$\inf_{\tau \in S(\Psi, G)} \rho_o(G, \tau) = \rho_o(G, T[\Psi, \rho_o, G]). \qquad (4.5)$$

Suppose there are s solutions of the asymptotic equations, so that $S(\Psi, F_{\theta_0}) = \{\theta_0, \dots, \theta_{s-1}\}$. We may have to restrict our parameter space in the following discussion to exclude infinite sets of solutions, for example, as are occasioned by going out well beyond a rejection point, for example. Define the ith functional root of Eq. (4.1) to be $T_i[\cdot]$, given by Eq. (4.5) with auxiliary functional $\rho(G, \tau) = |\tau - \theta_i|$. Suppose that for a parameter set D

$$\sup_{\tau \in D} \left| \int_R \Psi(x, \tau) \mathrm{d} F_n(x) - \int_R \Psi(x, \tau) \mathrm{d} F_{\theta_0}(x) \right| \to_{a.s} 0 \qquad (4.6)$$

and

$$\sup_{\tau \in D} \left| \int_R \left[\left\{ \frac{\partial}{\partial \tau} \Psi(x, \tau) \right\} dF_n(x) - \left\{ \frac{\partial}{\partial \tau} \Psi(x, \tau) \right\} dF_{\theta_0}(x) \right] \right| \to_{a.s} 0. \qquad (4.7)$$

We argue for the local consistency of solutions of equations (4.1) under Eqs. (4.6) and (4.7) using the arguments of Foutz (1977) but couched in the functional approach of Clarke (1983a) related in Chapter 2. Without loss of generality D is chosen to contain all the solutions of the asymptotic equations, that is $S(\Psi, F_{\theta_0}) \subset D$. The following theorem, which is given with permission, results.

Theorem 4.1: *(Clarke et al., 1993, Theorem 1) Assume Eqs. (4.6) and (4.7) hold. Then assume that $M(\theta_i)$, $i = 0, 1, \ldots, s - 1$, are non-singular matrices, where*

$$M(\theta_i) = \int_R \left\{ \frac{\partial}{\partial \tau} \Psi(x, \tau) \right\} \Big|_{\tau=\theta_i} dF_{\theta_0}(x). \qquad (4.8)$$

Let Ψ have continuous partial derivatives on $R \times D$, where D is some non-degenerate compact region containing $S(\Psi, F_{\theta_0})$ in its interior. Then there exist neighborhoods $N(T_i[F_{\theta_0}])$ of the parameters $T_i[F_{\theta_0}] = \theta_i$ such that for all sufficiently large n, $T_i[F_n]$ exists and is unique in $N(T_i[F_{\theta_0}])$, and $T_i[F_n]$ converges almost surely to $T_i[F_{\theta_0}]$. □

From Eqs. (4.6) and (4.7) it follows that for all sufficiently large n, $F_n \in n(\delta, F_{\theta_0})$, where the neighborhoods are defined in Eq. (2.17). The argument for consistency of $T_i[F_n]$ to $T_i[F_{\theta_0}]$ can be made analogous to the proof of Theorem 2.3 apart from condition A_0, which is replaced by $T[\psi, \rho_i, F_{\theta_0}] = \theta_i$. Assumptions $A_1 - A_4$ follow from the definition of the neighborhood Eq. (2.17) and the assumptions of Theorem 4.1. Hence, the assumptions of Theorem 2.3 are satisfied.

In essence we have shown $T_i[F_n] \to_{a.s} \theta_i$ for each of $i = 0, \ldots, s - 1$. There are also corresponding limit theorems, based on the asymptotic convergence of $n^{1/2}(T_i[F_n] - \theta_i)$ to a limiting normal distribution for each of $i = 0, \ldots, s - 1$, where the asymptotic variance depends on the ith asymptotic root θ_i of Eq. (4.2). We summarize this result, which is given with permission in the following.

Theorem 4.2: *(Clarke et al., 1993, Theorem 2) Assume the conditions of Theorem 4.1. Then each functional $T_i[\cdot]$ of Eqs. (4.3) and (4.5) satisfies a stochastic expansion*

$$|T_i[F_n] - T_i[F_{\theta_0}] - T'_{i, F_{\theta_0}} (F_n - F_{\theta_0})| = o_p \left(\frac{1}{n^{1/2}} \right), \qquad (4.9)$$

where

$$T'_{i, F_{\theta_0}} (F_n - F_{\theta_0}) = - \int_R M(\theta)^{-1} \Psi(x, \theta) d(F_n - F_{\theta_0})(x) |_{\theta=\theta_i}.$$

Moreover, $n^{1/2}(T_i[F_n] - T_i[F_{\theta_0}])$ converges in distribution as n tends to infinity to a random variable having a multivariate normal distribution with mean zero and dispersion matrix $\sigma^2(T_i, F_{\theta_0})$, where

$$\sigma^2(T_i, F_{\theta_0}) = M(\theta)^{-1} \int_R \Psi(x, \theta)\Psi^T(x, \theta)\mathrm{d}F_{\theta_0}(x)\{M(\theta)^{-1}\}^T|_{\theta=\theta_i}$$

□

The proof, detailed in Clarke et al. (1993), essentially uses the multivariate mean value theorem expansion of equations (4.1) at $T_i[F_n]$ about $T_i[F_{\theta_0}] = \theta_i$. Details are left for the reader's, investigation.

4.3   CONSISTENCY IN THE FACE OF MULTIPLE ROOTS

The problem of multiple roots to estimating equations often presents itself as a dilemma as to which root one should choose, and if having found a root, is it the right one? It is a problem that faces both applied and theoretical statisticians. Works and review papers by Heyde (1997), Heyde and Morton (1998), Gan and Jiang (1999), Small et al. (2000), and Small and Wang (2003) illustrate that examples abound in the literature, especially in regard to classical estimation vis-à-vis methods to do with maximum likelihood estimation. Biernacki (2005) expounds on the cornerstone work of Gan and Jiang (1999), giving a variation on the theme of those two authors. See also Blatt and Hero (2007) for a general discussion of asymptotics, including multivariate parameter likelihood estimation and testing.

The estimator is defined through not only the roots of the estimating equations but also by a "selection functional" used to choose the "consistent" root. Classical estimators such as the solutions of the method of moments equations are usually chosen to minimize some goodness of fit statistic in order to select a root in the event of multiple roots. A Cramér–von Mises statistic can be used as a goodness-of-fit statistic, for instance (Martin, 1936).

In this spirit, Gan and Jiang (1999) choose a solution of maximum likelihood equations where there are multiple roots by minimizing another "selection functional." Small and Wang (2003, section 5.9) quote the work of Heyde (1997) and Heyde and Morton (1998) where for estimating equations $g(\theta) = 0$ they suggest

1. *picking the root θ_i for which $\dot{g}(\theta_i)$ behaves asymptotically like $E_\theta \dot{g}(\theta)$ when evaluated at $\theta = \theta_i$ and*
2. *using a least squares or goodness-of-fit criterion to select the best root.*

Both 1. and 2. are encompassed in the theory of M-estimators in Clarke (1990) who gives general consistency arguments for the selected roots. See Corollary 2.2 and subsequent discussion in Section 2.2. We show here 1. is preferable to 2. in the case of Cauchy maximum likelihood equations, and indeed Theorem 4.3 establishes a weakly continuous and Fréchet differentiable solution of *1*. Other viewpoints on the Cauchy parametric family are found in Barnett 1966, Barnett and Lewis 1994, Clarke 1983a, pp. 1204–1205, Clarke 1986a, Copas 1975, Freedman and Diaconis 1982, Reeds 1985, Mizera 1996, Small and Yang 1999, Small and Wang 2003, and Freue 2007. Also 1. proves superior to 2. using Gan and Jiang's 1999 proposal, for the econometrics example from Amemiya (1994). This is established through Theorem 4.4 describing asymptotic normality of the statistics leading to the selection functionals, both at the true root θ and at alternative roots of the asymptotic estimating equation. The proof of Theorem 4.4 at θ for the special case of the Cauchy location parametric family affords a simple test of whether we have the correct root in that example. In another example, it is shown the method of Gan and Jiang (1999) works when *1*. does not. Hence, there is not an optimal approach available for all parametric families. Each parametric family and associated estimating equations must be considered separately. Also while *1*. and *2*. are proffered by Small and Wang (2003), following on from Heyde (1997) and Heyde and Morton (1998), importantly none of those authors offer a general consistency argument whereas at least for M-estimators under verifiable conditions quite some time before Clarke (1990) does.

4.3.1 Preliminaries

Returning to the selection functional one can imagine that if ρ_o was chosen to be an objective function that was minimized over the parameter space Θ and that $S(\Psi, G)$ was a set of stationary points formed by setting $K_G(\tau)$ as the set of partial derivatives of ρ_o with respect to the parameter variables, then we would have the classical estimation problems associated with likelihood theory. For example, here maximizing the log-likelihood is the same as minimizing the negative of the log-likelihood which is done by setting

$$\rho_o(G, \tau) = -\int_R \log f_\tau(x) dG(x), \tag{4.10}$$

assuming f_τ is the density function corresponding to the distribution F_τ and that Ψ is the efficient score function. That is, when $G \equiv F_n$ we effectively maximize

$$\mathcal{L} = \sum_{j=1}^{n} \log f_\tau(X_j). \tag{4.11}$$

Examples of selection functionals that need not be directly related to the estimating equations include Hampel's choice, $\rho_o(G, \tau) = |G^{-1}(\frac{1}{2}) - \tau|$, which defines the root closest to the median. The method of moments estimator, for example for data defined on $\mathcal{E}$ with finite rth moment, can be defined through a choice of $\Psi(x, \theta) = (x - E_\theta[X], \ldots, x^r - E_\theta[X^r])^T$, where E_θ represents the expectation operator with respect to the parametric distribution, and in the event of more than one solution to either Eq. (4.1) or Eq. (4.3), one can use a goodness-of-fit criterion such as the Cramér–von Mises distance for ρ_o to select the consistent root.

The attractive though at that time tentative proposal for a robust selection functional given in Clarke (1990, p. 154) is to let

$$\rho_o(F_n, \tau) = \left| \int \left\{ \frac{\partial}{\partial \tau} \Psi(x, \tau) \right\} dF_\tau(x) - \int \left\{ \frac{\partial}{\partial \tau} \Psi(x, \tau) \right\} dF_n(x) \right| \qquad (4.12)$$

Essentially, this is based on uniform convergence theory, which is in the form of almost sure ($a.s.$) convergence where assuming $\theta \equiv \theta_o$:

$$\sup_{\tau \in \Theta} |K_{F_n}(\tau) - K_{F_\theta}(\tau)| \to_{a.s.} 0, \qquad (4.13)$$

and

$$\sup_{\tau \in \Theta} \left| \frac{\partial}{\partial \tau} K_{F_n}(\tau) - \frac{\partial}{\partial \tau} K_{F_\theta}(\tau) \right| \to_{a.s.} 0. \qquad (4.14)$$

These are essentially Eqs. (4.6) and (4.7) with D replaced by the potentially larger set Θ.

Using a selection functional to identify an asymptotically unique consistent root of equations (4.1) presumes there exists such a root, which rightly requires conditions on the function $\Psi(\cdot, \cdot)$. These are afforded by conditions **A** and/or conditions **A'**. Results do not appear to extend to discontinuous forms of Ψ.

The Likelihood Approach of Gan and Jiang
GJ based on earlier work by White (1982) point out under regularity conditions that

$$E_\theta \left[\frac{\partial}{\partial \tau} \mathcal{L}_{|_{\tau=\theta}} \right] = 0 \qquad (4.15)$$

and also

$$E_\theta \left[\left\{ \frac{\partial}{\partial \tau} \mathcal{L} \right\} \left\{ \frac{\partial}{\partial \tau} \mathcal{L} \right\}^T {}_{|_{\tau=\theta}} \right] + E_\theta \left[\frac{\partial^2}{\partial \tau^2} \mathcal{L}_{|_{\tau=\theta}} \right] = 0. \qquad (4.16)$$

Equation (4.15) is consistent with the likelihood equation

$$\frac{\partial}{\partial \tau} \mathcal{L} = 0.$$

GJ claim that a global maximizer of the likelihood would satisfy

$$\left\{\frac{\partial}{\partial\tau}\mathcal{L}\right\}\left\{\frac{\partial}{\partial\tau}\mathcal{L}\right\}^{T} + \frac{\partial^2}{\partial\tau^2}\mathcal{L} \approx 0, \tag{4.17}$$

whereas an inconsistent root would not. *GJ* restrict their view to a univariate parameter space in their discussion, though this extends in a natural way to vector parameters along the lines of White (1982). For example, *GJ* consider two ways of computing the asymptotic "standard errors" of the estimated parameters θ:

$$SD_1 = \left[E_\theta\left\{\left(\frac{\partial}{\partial\tau}\log f_\tau(X)|_{\tau=\theta}\right)^2\right\}\right]^{-\frac{1}{2}}. \tag{4.18}$$

This is denoted the outer product form. The alternative way is the Hessian form which is

$$SD_2 = \left[-E_\theta\left\{\frac{\partial^2}{\partial\tau^2}\log f_\tau(X)|_{\tau=\theta}\right\}\right]^{-\frac{1}{2}}. \tag{4.19}$$

Visualize, SD_1^{-2} and SD_2^{-2} both give Fisher information which we denote $\mathcal{I}(\theta)$ and the asymptotic variance of the estimated parameters (at the parametric model) is the inverse of Fisher information. The test statistic proposed by White as a goodness of fit test, but essentially employed by *GJ* as a particular form to select the roots of the likelihood equations that best fit is then

$$\phi_2(\tau) = \frac{1}{n}\sum_{j=1}^{n}\left[\left\{\frac{\partial}{\partial\tau}\log f_\tau(X_j)\right\}^2 + \frac{\partial^2}{\partial\tau^2}\log f_\tau(X_j)\right] \approx \{\hat{SD}_1^{-2} - \hat{SD}_2^{-2}\}. \tag{4.20}$$

The argument is that for the solution of best fit $\phi_2(\hat\theta) \approx 0$. To put this in the framework of the selection functional approach, let the choice of the efficient score function be

$$\Psi(x,\tau) = \frac{\partial}{\partial\tau}\log f_\tau(x) = \frac{\frac{\partial}{\partial\tau}f_\tau(x)}{f_\tau(x)}$$

in the same vein as the more general M-estimation method, whereupon the goodness-of-fit statistic suffices as a selection functional, where

$$\rho_o(F_n,\tau) = |\phi_2(\tau)| = \left|\int\Psi(x,\tau)\Psi(x,\tau)^T dF_n(x) + \int\frac{\partial}{\partial\tau}\Psi(x,\tau)dF_n(x)\right|. \tag{4.21}$$

The question arises does this quantity appear to have the qualities of a selection functional that would distinguish the consistent root from among multiple roots of the estimating equations? Sufficient conditions in order that a selection functional facilitates the consistency argument (indeed in terms of establishing a weakly continuous root that is selected from the set $S(\Psi, G)$) are given in Formulae (4.1) and (4.2) in Clarke (1983a, p. 1200), related in Formulae (2.12) and (2.13) of this book. For ease of reference, Formula (2.12) is repeated as follows:

$$\forall \text{ open neighborhood } N \text{ of } \theta, \quad \inf_{\tau \notin N} \rho_o(F_\theta, \tau) - \rho_o(F_\theta, \theta) > 0 \qquad (4.22)$$

It can be noted that this condition can preclude lack of identifiability of the parametric model since if there is lack of identifiability there are at least two distinct parameters $\theta_1 \neq \theta_2$ for which $F_{\theta_1} = F_{\theta_2}$, and we could find an open neighborhood of say θ_1 not containing θ_2 for which $\rho_o(F_{\theta_1}, \theta_2) - \rho_o(F_{\theta_1}, \theta_1) = 0$. This is particularly the case with distance metrics where $\rho_o(F_\theta, \theta) = 0$, for instance. See also the discussion in Clarke (1990, p. 151) for likelihood theory and see Clarke et al. (1993) and also the data example in Clarke (2000b) for just two examples involving potential lack of identifiability problems.

Example 4.1: Consider the estimation of location of a Cauchy distribution with scale known. The parametric density is $f_\theta(x) = \frac{1}{\pi(1+(x-\theta)^2)}$, where $-\infty < x < \infty$. Here it is not hard to establish that the selection functional of GJ at the Cauchy model corresponds to

$$\rho_o(F_\theta, \tau) = |\int \frac{6(x-\tau)^2 - 2}{[1+(x-\tau)^2]^2 \pi[1+(x-\theta)^2]} dx|. \qquad (4.23)$$

It is easily shown $\rho_o(F_\theta, \theta) = 0$. While this is a unique minimum, it does not satisfy Eq. (4.22) since it can be observed that both $\lim_{\tau \to +\infty} \rho_o(F_\theta, \tau)$ and also $\lim_{\tau \to -\infty} \rho_o(F_\theta, \tau)$ are 0. This suggests that, empirically, "erroneous" roots of the likelihood equations far from the bulk of the data have potential to be "selected" by such a selection functional. Compare with the discussion of Small et al. (2000, section 2.2). On the other hand, consider the selection functional of Clarke (1990) in Eq. (4.12) for this example. This in fact is the same proposal as formulated in point *1.* in the summary of Small and Wang (2003) of the work of Heyde (1997) and Heyde and Morton (1998). Here we find the selection functional has the form

$$\rho_o(F_\theta, \tau) = |-\frac{1}{2} - \int \frac{-2 + 2(x-\tau)^2}{(1+(x-\tau)^2)^2} \frac{1}{\pi(1+(x-\theta)^2)} dx|. \qquad (4.24)$$

This selection functional has as its limit one half as τ tends to plus and minus infinity. Indeed, it is possible with the aid of the symbolic mathematics facility in MATLAB to plot both the selection functionals evaluated at the Cauchy model on the same graph when say we set without loss of generality $\theta = 0$. Figure 4.1 illustrates that the selection functional of Eq. (4.24) is superior in the sense that

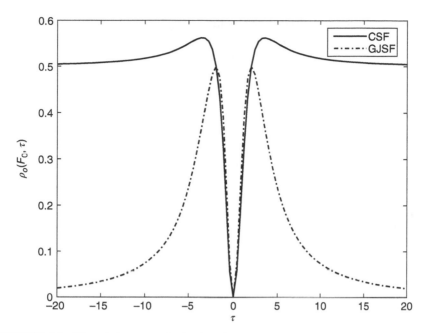

FIGURE 4.1   Plot of the mooted selection functional of Clarke (1990) (CSF) and the selection functional of Gan and Jiang (1999)(GJSF) at the Cauchy model. (*Source*: Gan and Jiang (1999)).

it does not redescend to zero as τ tends to infinity, while on the other hand, it behaves like the selection functional of *GJ*, formula (4.23), in the vicinity of the true parameter.   □

Other Choices for Selection Functional

Of course if we wished to tread down the path of maximum likelihood estimation there is some doubt about the existence of the selection functional (4.10) for general G, yet as noted by Mizera (1996), an alternative selection functional which does exist under mild conditions and which provides the same estimating functional, i.e. the maximum likelihood estimator, is given by

$$\rho_o(G, \tau) = \int \log \frac{(1 + (x - \tau)^2)}{1 + x^2} dG(x) \qquad (4.25)$$

Unfortunately, this function is not easily plotted at say $G = F_\theta$ which gives us little indication as to how it may behave in comparison to the other two selection functionals, albeit the selected roots are weakly continuous at F_θ.

Further work improving on the method of *GJ* is given in Biernacki (2005). There the criterion for selecting a root is based on an asymptotic result that if $\hat{\theta}_n$ is the consistent root to θ_o, then letting $v(\theta_o) = \text{var}_{\theta_o}[\log f(X, \theta_o)]$, and also setting

$$\phi_2^B(\theta) = \int \log f_\theta(x) dF_n(x) - \int \log f_\theta(x) dF_\theta(x)$$

then,

$$\frac{\sqrt{n}\phi_2^B(\hat{\theta}_n)}{v(\hat{\theta}_n)} \to_d N(0, 1). \tag{4.26}$$

In essence, it is therefore implied that one choose a root according to

$$\text{argmin} \frac{|\phi_2^B(\hat{\theta}_n)|}{v(\hat{\theta}_n)}.$$

Now since for a location model $v(\theta_o)$ does not depend on θ_o, the selection functional would in this instance depend only on

$$\rho_o(F_n, \tau) = | \int \{\log f_\tau(x) - E_o[\log f_o(X)]\} dF_n(x)|,$$

which albeit works empirically when the empirical distribution F_n is guaranteed to be from the chosen parametric family, essentially choosing a root so that it minimizes a generalized moment condition, there are clearly going to be robustness issues for F_n generated from G near in some sense but not exactly equal to some F_{θ_o}. This is because the function $\log f_\tau(x)$ is unbounded as a function of the observation space variable whenever the parametric density is spread across the whole real line say as is also in the case for the Cauchy parametric family.

Robustness in the Case of the Mooted Selection Functional

The second sufficient condition to be satisfied by a selection functional in order that the functional be weakly continuous is Eq. (4.2) in Clarke (1983a) or Formula (2.13) in this book. Here we state it in terms of the Prohorov distance between distributions, d_P, since continuity with respect to the Prohorov metric of the resulting estimating functional leads to weak continuity of such a functional with all the important consequences–strong consistency and qualitative robustness–in the sense of Hampel (1971), see also Huber and Ronchetti (2009). The condition is then as follows:

$$\forall \eta > 0 \text{ there exists an } \quad \epsilon > 0 \text{ such that } d_P(G, F_\theta) < \epsilon$$

$$\text{implies} \quad \sup_{\tau \in \Theta} |\rho_o(G, \tau) - \rho_o(F_\theta, \tau)| < \eta$$

Fortunately, in the case of the selection functional in Eq. (4.12), this result can be established at the Cauchy distribution F_θ using Theorem 2.1. Thus, we have the following result.

Theorem 4.3: *For the Cauchy location parametric family let ψ be the efficient score function and let the selection functional, $\rho_o(G, \tau)$, be given by the form mooted in Clarke (1990), more specifically Formula (4.24). Then the estimating functional $T[\psi, \rho_o, \cdot]$ is weakly continuous at F_θ. In addition the functional is Fréchet differentiable at F_θ with respect to the Kolmogorov metric d_k. The functional is also fully efficient at the Cauchy model.*

Proof. It is enough to establish the conditions **A** in the special case of the efficient score function ψ for the Cauchy location model. For the Cauchy density $\psi(x, \tau) = \frac{2(x-\tau)}{1+(x-\tau)^2}$ and $\frac{\partial}{\partial \tau}\psi(x, \tau) = \frac{-2+2(x-\tau)^2}{[1+(x-\tau)^2]^2}$. Condition A_1 is clearly satisfied. Now, in fact $\sup_{\tau \in \mathcal{E}}|\frac{\partial}{\partial x}\psi(x, \tau)| \leq 2$. Hence, by the mean value theorem $\sup_{\tau \in \Theta}|\psi(x, \tau) - \psi(y, \tau)| \leq 2|x - y|$. This Lipschitz result is strong and implies the weaker concept of equicontinuity of the class of functions $\mathcal{A}_1 = \{\psi(x, \tau) : \tau \in \mathcal{E}\}$. This class of functions is also clearly uniformly bounded.

Also it is possible, with the aid of a symbolic maths package, to establish $\sup_{\tau \in \mathcal{E}}|\frac{\partial}{\partial x}\frac{\partial}{\partial \tau}\psi(x, \tau)| \leq \sqrt{2} + \frac{3}{2}$. The function $\frac{\partial}{\partial \tau}\psi(x, \tau)$ is also uniformly bounded. Hence, it follows the class of functions $\mathcal{A}_2 = \{\frac{\partial}{\partial \tau}\psi(x, \tau) : \tau \in \mathcal{E}\}$ is also equicontinuous and uniformly bounded.

Conditions A_0, A_1, and A_2 are clearly satisfied. Condition A_3 is satisfied by the fact that Fisher information $= 1/2$ (Hampel et al., 1986, p. 104). Condition A_4 is satisfied by appealing to Theorem 2.1 for both classes of functions $\mathcal{A}_1$ and $\mathcal{A}_2$. It has also been illustrated that Eq. (2.12) is satisfied for the choice Eq. (4.12) and since the latter involves partial derivatives of ψ with respect to the parameter, Theorem 2.1 can be invoked again to establish Eq. (2.13). Weak continuity of the statistical functional $T[\psi, \rho_o, .]$ follows from Theorem 2.4 and Fréchet differentiability follows from Theorem 2.6 and Corollary 2.4. Again note that Fréchet differentiability has automatic consequences in terms of asymptotic normality of the resulting estimator and in this case asymptotic efficiency of the estimator since the asymptotic distribution is governed by the efficient score function alone, the selection functional having guided us to the efficient root.

Remark 4.1: It is but a small but laborious step to strengthen the argument to say that $T[\psi, \rho_o, .]$ is both weakly continuous and Fréchet differentiable at each F in a neighborhood of F_θ (Clarke, 2000a).

The Example from Amemiya

Example 4.2: We illustrate here classical consistency arguments, that is, arguments for consistency of some non-robust estimating functionals by considering Example 4 of *GJ* which is taken from Amemiya (1994). Here we assume $X_1 \sim N(\theta, \theta^2)$. Then the efficient score function is

$$\psi(x, \tau) = \frac{\frac{\partial}{\partial \tau} f_\tau(x)}{f_\tau(x)} = \frac{-1}{\tau} + \frac{x^2}{\tau^3} - \frac{x}{\tau^2},$$

so that

$$K_{F_n}(\tau) = \frac{-1}{\tau} + \frac{\frac{1}{n} \sum_{j=1}^n X_j^2}{\tau^3} - \frac{\frac{1}{n} \sum_{j=1}^n X_j}{\tau^2}$$

and from the strong law of large numbers and due to existence of moments we gain the asymptotic curve

$$K_{F_\theta}(\tau) = E_\theta[\psi(X, \tau)] = \frac{-1}{\tau} + \frac{2\theta^2}{\tau^3} - \frac{\theta}{\tau^2}.$$

This is $d_1(\tau)$ in notation of *GJ*. To illustrate say the uniform convergence such as in Eq. (4.13), consider a parameter space that is restricted to be away from zero, that is where $\Theta = (-\infty, -\delta] \cup [\delta, \infty)$ for some small $0 < \delta < 1$. Then

$$\sup_{\tau \in \Theta} |K_{F_n}(\tau) - K_{F_\theta}(\tau)| \leq \sup_{\tau \in \Theta} \left\{ \frac{|\frac{1}{n} \sum_{j=1}^n X_j^2 - 2\theta^2|}{\tau^3} + \frac{|\frac{1}{n} \sum_{j=1}^n X_j - \theta|}{\tau^2} \right\}$$

$$\leq \frac{1}{\delta^3} \left\{ |\frac{1}{n} \sum_{j=1}^n X_j^2 - E_\theta[X^2]| + |\frac{1}{n} \sum_{j=1}^n X_j - E_\theta[X]| \right\}$$

$$\rightarrow_{a.s.} 0 \quad \text{(by the strong law of large numbers)}$$

Similar arguments which we do not give here lead to Eqs. (4.14) and (4.27), all of which depend on the existence of moments for this parametric density. As noted in *GJ*, there are two possible solutions to the equation $K_{F_\theta}(\tau) = 0$, they being θ and also -2θ. The selection functional of *GJ* evaluated at its asymptotic limit is in essence (after some calculation)

$$\rho_o(F_\theta, \tau) = |\frac{2}{\tau^2} + \frac{4\theta}{\tau^3} - \frac{8\theta^2}{\tau^4} - \frac{8\theta^3}{\tau^5} + \frac{10\theta^4}{\tau^6}|$$

This is in fact $d_2(\tau)$ of Gan and Jiang (1999). It is easily seen that $\rho_o(F_\theta, \theta) = 0$. In Figure 4.2 this corresponds to the zero of the dashed line at $\tau = \theta = 1$. However, there are other possible values of τ at which the selection functional of *GJ* is zero, which are in fact $\{1.2143\theta, -2.6751\theta, -1.5392\theta\}$. There are four solutions corresponding to a polynomial of degree four. It is worthy of note that this selection functional has two zeros at the asymptotic curve which are both not that far from

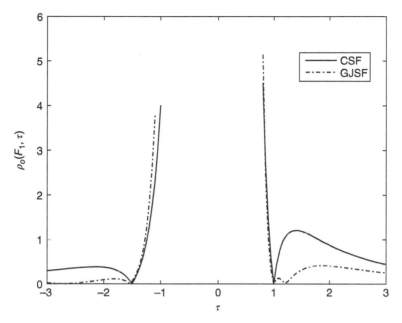

FIGURE 4.2 Plot of the mooted selection functional of Clarke (1990) (CSF) and the selection functional of Gan and Jiang (GJSF) for Example 4 of Gan and Jiang (1999). (*Source*: Gan and Jiang (1999)).

the erroneous root of -2θ. On the other hand, the selection functional in Eq. (4.12) has a simple form, namely

$$\rho_o(F_\theta, \tau) = |\frac{-4}{\tau^2} + \frac{-2\theta}{\tau^3} + \frac{6\theta^2}{\tau^4}|$$

Again note that $\rho_o(F_\theta, \theta) = 0$ which in Figure 4.2 corresponds to the zero at $\tau = 1$. There is only one other zero of the mooted selection functional at $\tau = -1.5\theta$, these two solutions corresponding to roots of a quadratic. The advantage that the selection functional of Eq. (4.12) has over that from *GJ*, see Figure 4.2, is that in the vicinity of the erroneous root of the asymptotic likelihood equations, that is near $-2\theta = -2$, the mooted selection functional is greater than that of *GJ*. In fact, the value of the selection functional of *GJ* at the distribution $F_\theta = F_1$ and at the erroneous root of $-2\theta = -2$ is $3/32\theta^2 = 0.093\ 75/\theta^2$, whereas that of the selection functional of Clarke (1990) given here in Formula (4.12) is $3/8\theta^2 = 0.375/\theta^2$. A more general value of θ other than $\theta = 1$ used for the illustration can be substituted. (Note there is an asymptote at $\tau = 0$.)   □

Example 4.3: The form of the selection functional (4.12) does not always work for every model. Consider the case of $X_1 \sim$ Poisson(e^θ). Here the efficient score function is $\psi(x, \tau) = x - e^\tau$. Indeed we find here that the partial derivative of $\frac{\partial}{\partial \tau}\psi(x, \tau) = -e^\tau$ which is a constant with respect to the observation space variable, whereupon the form Eq. (4.12) is uniformly zero! On the other hand, the selection functional of GJ is able to discriminate between the two roots of the asymptotic equation.   □

4.3.2   Asymptotic Properties of Roots and Tests

GJ following the theory of White (1982) provide in Theorem 2 of their paper an asymptotic normality result, first for the asymptotically consistent and efficient root to the true parameter and also simultaneously of a test statistic (Gan and Jiang, 1999, formula (16)), that may be used to test whether one has the efficient root, in order that one may cease searching for further roots. In this section, we generalize the test statistics to consider the "efficient" root using the selection functionals of either Clarke (1990) or that of GJ. Here "efficient" root can refer to the asymptotically unique consistent root of an M-estimating equation which may not be fully efficient having asymptotic variance different to the inverse of Fisher Information but can have favorable qualities from the point of view of infinitesimal properties such as described by Hampel's influence function (Hampel et al. 1986). Also see Remark 4.2. Further the asymptotic normality of subsequent test statistics at roots other than the efficient root is considered. This say allows us to derive the power of the test statistics when at a root other than the efficient root, something which is not considered previously, other than by using simulation. We then illustrate this using the Example 4.2 in the following section.

Theorem 4.4: *Assume conditions (4.13) and (4.14) hold. Also assume condition (4.1) of Clarke (1990) holds, that is, in the sense that*

$$\sup_{\tau \in \Theta} |\rho_o(F_n, \tau) - \rho_o(F_\theta, \tau)| \to_{a.s.} 0. \qquad (4.27)$$

Let $\{\theta_i, i = 0, \dots, s-1\}$ be s roots of the asymptotic equations $K_{F_\theta}(\theta_i) = 0$. For example, $\theta_0 \equiv \theta$ in the notation of this book. Then assuming $M_\theta(\theta_i) = \int_R \{\frac{\partial}{\partial \tau}\psi(x, \tau)\}dF_\theta(x)|_{\tau=\theta_i} \neq 0$, there exists unique consistent roots of Eq. (4.1), $\hat{\theta}_{in}$ for θ_i for which $\sqrt{n}(\hat{\theta}_{in} - \theta_i)$ converges in distribution to a normal random variable with mean zero and variance

$$\sigma_\theta^2(\theta_i) = \int_R \psi(x, \tau)^2 dF_\theta(x)/M_\theta(\theta_i)^2|_{\tau=\theta_i},$$

for each $i = 0, 1, \ldots, s - 1$. In addition let $\mu_\theta(\theta_i) = E_\theta[\tilde{\psi}(X, \theta_i)]$. (Examples of $\tilde{\psi}$ are given in the next section.) Then

$$\frac{1}{\sqrt{n}} \sum_{j=1}^{n} \tilde{\psi}(X_j, \hat{\theta}_{in}) \to_d N\left(\sqrt{n}\mu_\theta(\theta_i), \left\{ \frac{J_\theta(\theta_i)}{M_\theta(\theta_i)} \right\}^2 \sigma_{\psi_2}^2(\theta, \theta_i) \right). \tag{4.28}$$

Here $\sigma_{\psi_2}^2(\theta, \theta_i) = \text{var}_\theta[\psi_2(X, \theta_i)]$ and

$$\psi_2(x, \tau) = [-\psi(x, \tau) + \{M_\theta(\tau)J_\theta(\tau)^{-1}\}\tilde{\psi}(x, \tau)] \tag{4.29}$$

and $J_\theta(\tau) = E_\theta[\frac{\partial}{\partial \tau}\tilde{\psi}(X, \tau)]$.

Proof. To each root θ_i of the asymptotic equation $K_{F_\theta}(\theta_i) = 0$, there is a unique consistent sequence of solutions $\hat{\theta}_{in}$ to θ_i. See Theorem 4.1. (Here θ may be a vector parameter.) Asymptotic normality of $\sqrt{n}(\hat{\theta}_{in} - \theta_i)$ is proved in Theorem 4.2. To illustrate asymptotic normality of the test statistic (4.28), we revert to a univariate parameter and use the mean value theorem,

$$\frac{1}{\sqrt{n}} \sum_{j=1}^{n} \tilde{\psi}(X_j, \hat{\theta}_{in}) = \frac{1}{\sqrt{n}} \sum_{j=1}^{n} \tilde{\psi}(X_j, \theta_i)$$

$$+ \left\{ \frac{\partial}{\partial \theta} \frac{1}{n} \sum_{j=1}^{n} \tilde{\psi}(X_j, \theta) \right\}_{|\theta - \theta_{in}^*} \sqrt{n}(\hat{\theta}_{in} - \theta_i), \tag{4.30}$$

where θ_{in}^* is between $\hat{\theta}_{in}$ and θ_i. Then by uniform convergence,

$$\left\{ \frac{1}{n} \sum_{j=1}^{n} \frac{\partial}{\partial \theta} \tilde{\psi}(X_j, \theta) \right\}_{|\theta = \theta_{in}^*} \to_{a.s} J_\theta(\theta_i).$$

Also, again by the mean value theorem

$$0 = K_{F_n}(\hat{\theta}_{in}) = K_{F_n}(\theta_i) + \frac{\partial}{\partial \theta} K_{F_n}(\theta)_{|\theta=\tilde{\theta}_{in}} (\hat{\theta}_{in} - \theta_i), \tag{4.31}$$

where $\tilde{\theta}_{in}$ is between $\hat{\theta}_{in}$ and θ_i, and uniform convergence gives

$$\frac{\partial}{\partial \theta} K_{F_n}(\theta)_{|\theta=\hat{\theta}_{in}} \to_{a.s.} M_\theta(\theta_i),$$

hence from Eq. (4.31) we have

$$\sqrt{n}(\hat{\theta}_{in} - \theta) \approx -M_\theta(\theta_i)^{-1} \frac{1}{\sqrt{n}} \sum_{j=1}^{n} \psi(X_j, \theta_i). \tag{4.32}$$

Substituting this into Eq. (4.30)

$$
\frac{1}{\sqrt{n}} \sum_{j=1}^{n} \tilde{\psi}(X_j, \hat{\theta}_{\text{in}})
$$

$$
\approx \frac{1}{\sqrt{n}} \sum_{j=1}^{n} -J_\theta(\theta_i) M_\theta(\theta_i)^{-1} \psi(X_j, \theta_i) + \frac{1}{\sqrt{n}} \sum_{j=1}^{n} \tilde{\psi}(X_j, \theta_i)
$$

$$
= J_\theta(\theta_i) M_\theta(\theta_i)^{-1} \frac{1}{\sqrt{n}} \sum_{j=1}^{n} \{ M_\theta(\theta_i) J_\theta(\theta_i)^{-1} \tilde{\psi}(X_j, \theta_i) - \psi(X_j, \theta_i) \}
$$

$$
\to_d N \left(\sqrt{n} \mu_\theta(\theta_i), \left(\frac{J_\theta(\theta_i)}{M_\theta(\theta_i)} \right)^2 \sigma_{\psi_2}^2(\theta, \theta_i) \right)
$$

since it is a scaled normed sum of independent normal variables each with mean $\mu_\theta(\theta_i)$. Implicit in the above argument is use of strong almost sure convergence implies convergence in probability whereupon use of Slutsky's Lemma gives the appropriate convergence in distribution of the test statistic.

Remark 4.2: Theorem 4.4 is a broader result than the one proffered by *GJ*, which seeks only the distribution of the test statistic at the asymptotically unique consistent root for the maximum likelihood equations to θ_0 that is described in their formula (14). The notation used in Theorem 4.4 is consistent with the definition of ψ used in M-estimation and is necessarily different from the notation in *GJ*. Theorem 4.4 generalizes the results to include s asymptotically unique consistent roots to s zeros of the asymptotic equations.

Remark 4.3: Although for brevity for Theorem 4.4 we have assumed that θ is a real valued parameter, there is no essential difficulty to generalize the results to the case, where θ is multidimensional. See Blatt and Hero (2007) for likelihood theory extensions. See also the conclusion in *GJ*.

4.3.3 Application of Asymptotic Theory

The detailed and somewhat intricate calculations for the four-numbered formulae in this section can be found in Appendix B which also relies on Appendix C. These then lead to an elementary comparison of power in our second example which we address first. On the other hand, the asymptotic theory in the case of the first example leads to a useful and elementary test for whether a root is a consistent one in the case of the Cauchy family of distributions.

Example 4.2 (continued): *GJ* attempt through Monte Carlo simulation, the results of which they exhibit in their Table 3, to illustrate the performance of their test statistic in the example from Amemiya and essentially find very large samples are needed to retain the level of the test. They also comment that "The test suffers the problem of over-rejection at relatively smaller samples. The small-sample property of this test certainly deserves further investigation." It is worth, therefore, trying to fathom the asymptotic theory associated with this test, for both the choice of selection functional proposed by *GJ* and also the selection functional (4.12) proposed initially in Clarke (1990). Recall, for the choice of ψ being the efficient score function, *GJ* choose say

$$\tilde{\psi}(x, \tau) = \psi^2(x, \tau) + \frac{\partial}{\partial \tau} \psi(x, \tau),$$

whereas the choice of proposal in Clarke (1990) is say

$$\tilde{\psi}(x, \tau) = -\frac{\partial}{\partial \tau} \psi(x, \tau) + \int \left(\frac{\partial}{\partial \kappa} \psi(x, \kappa) \right) dF_\kappa(x)\big|_{\kappa = \tau}.$$

In each case, respectively, the selection functional is

$$\rho_o(G, \tau) = \left| \int \tilde{\psi}(x, \tau) dG(x) \right|.$$

The following can be established from the asymptotic theory: there are asymptotically two unique consistent roots of the likelihood equations, these being say $\{\hat{\theta}_{on}\}$ and $\{\hat{\theta}_{1n}\}$ each of which is consistent to θ and -2θ respectively. After some calculations, for the statistic involved implicitly to select the correct root in the choice of Clarke's recommended selection functional we have

$$\frac{1}{\sqrt{n}} \sum_{j=1}^{n} \tilde{\psi}_{Cl}(X_j, \hat{\theta}_{on}) \to_d N\left(0, \frac{2}{3\theta^4}\right). \tag{4.33}$$

On the other hand, we find that for *GJ*'s choice of selection functional that

$$\frac{1}{\sqrt{n}} \sum_{j=1}^{n} \tilde{\psi}_{GJ}(X_j, \hat{\theta}_{on}) \to_d N\left(0, \frac{152}{3\theta^4}\right). \tag{4.34}$$

Similar calculations as a result of Theorem 4.4 can be used to derive the distribution of the statistic at the incorrect root, for example, using Clarke's recommended selection functional we find

$$\frac{1}{\sqrt{n}} \sum_{j=1}^{n} \tilde{\psi}_{Cl}(X_j, \hat{\theta}_{1n}) \to_d N\left(\frac{-3\sqrt{n}}{8\theta^2}, \frac{97}{384\theta^4}\right), \tag{4.35}$$

whereas for the selection functional of *GJ* at the erroneous consistent root (to -2θ) we find

$$\frac{1}{\sqrt{n}} \sum_{j=1}^{n} \tilde{\psi}_{GJ}(X_j, \hat{\theta}_{1n}) \to_d N\left(\frac{-3\sqrt{n}}{32\theta^2}, \frac{73}{384\theta^4}\right). \tag{4.36}$$

Now supposing the true $\theta = 1$ and we had the root $\hat{\theta}_{1n}$ to test. Recall $\hat{\theta}_{1n}$ converges to $-2\theta = -2$. Assuming we have a consistent estimate of the standard deviation of our test statistic, by bootstrapping for example. (See Section 5.1.3 for a discussion of the bootstrap.) For instance, as suggested by GJ one may set as a test statistic $T_n/\text{SD}(T_n)$, where $T_n = \frac{1}{\sqrt{n}} \sum \tilde{\psi}(X_j, \hat{\theta})$, which has an asymptotic $N(0, 1)$ distribution, when bootstrapping with the correct root. On the other hand, the power of our test can be evaluated with knowledge of the asymptotic alternative distribution given in either Eq. (4.35) or Eq. (4.36), respectively, for example. Thus, the asymptotic power for the statistic of GJ, for example for a test carried out at the 5% significance level, is calculated using Eq. (4.36) as
$P\left(Z < -1.96 + \frac{3\sqrt{n}}{32}/\sqrt{\frac{73}{384}}\right) + P\left(Z > 1.96 + \frac{3\sqrt{n}}{32}/\sqrt{\frac{73}{384}}\right)$. Here Z is a standard normal variable. The power calculations in Table 4.1 show the test as a result of choosing the mooted selection functional of Clarke (1990) as an alternative to GJ clearly wins. As a remark the simulations of GJ on the power of the statistic give empirical values which generally agree with the asymptotic values above; values from their table 3 are reported as empirical power. Empirical values of the size of the test using the selection functional mooted by Clarke are also important. Assuming the true value $\theta = 1$ the size of the test are given in Table 4.2. The size of the number of bootstrap samples was taken to be 1 000 here. *Example* 4.1(continued): The first example of this section is that of the location only Cauchy parametric family. Here there is only one unique consistent root, $\hat{\theta}_{on}$, to the asymptotic likelihood equations which converges almost surely to the asymptotic root θ. Also there can be spurious roots that can be thrown up by the empirical equations, none of which are consistent to an alternative to θ. These are as outlined in such works as Freedman and Diaconis (1982). The asymptotic distribution of Clarke's recommended selection functional is, however, not governed by a straightforward application of Theorem 4.4 where $\theta_i = \theta$ for the following reason. It can be observed for this parametric family and the choice of the selection functional mooted by Clarke that $J_\theta(\theta) = 0$. All is not lost however, since appealing to expansion (4.30), and noting there is no contribution from the second term of the right-hand side, then

$$T_n^{\text{Cl}} = \frac{1}{\sqrt{n}} \sum_{j=1}^n \tilde{\psi}(X_j, \hat{\theta}_{on}) \approx \frac{1}{\sqrt{n}} \sum_{j=1}^n \tilde{\psi}(X_j, \theta)$$

$$\rightarrow_d N(E_\theta[\tilde{\psi}(X, \theta)], \text{var}_\theta[\tilde{\psi}(X, \theta)])$$

The latter distribution is easily evaluated using symbolic algebra when ψ corresponds to the efficient score function and is in fact $N\left(0, \frac{17}{8}\right)$. Hence, we would not reject $\hat{\theta}$ as being the efficient root if indeed $T_n^{\text{Cl}} \in \left(-1.96 \times \sqrt{\frac{17}{8}}, 1.96 \times \sqrt{\frac{17}{8}}\right) = (-2.857, 2.857)$ for a test at the 5% significance level.   □

TABLE 4.1   Power Calculations of the Tests

| | \multicolumn{8}{c}{$\alpha = 0.05$} |
| | | | | | | | |

| | \multicolumn{8}{c}{n} |

	10	15	20	40	100	250	500	1000
Power T_n^{Cl}	0.655	0.824	0.916	0.997	1.000	1.000	1.000	1.000
Power T_n^{GJ}	0.104	0.132	0.161	0.275	0.575	0.925	0.998	1.000
Emp. Power T_n^{GJ}	na	na	na	na	na	0.866	0.980	1.000

| | \multicolumn{8}{c}{$\alpha = 0.10$} |

| | \multicolumn{8}{c}{n} |

	10	15	20	40	100	250	500	1000
Power T_n^{Cl}	0.763	0.893	0.955	0.999	1.000	1.000	1.000	1.000
Power T_n^{GJ}	0.177	0.215	0.252	0.389	0.693	0.960	0.999	1.000
Emp. Power T_n^{GJ}	na	na	na	na	na	0.908	0.998	1.000

TABLE 4.2   Global Empirical Size Calculations of the Test using T_n^{Cl}

| | \multicolumn{8}{c}{n} |

	10	15	20	40	100	250	500	1000
$\alpha = 0.05$	0.117	0.106	0.096	0.068	0.070	0.047	0.050	0.054
$\alpha = 0.10$	0.151	0.150	0.126	0.111	0.121	0.090	0.100	0.100

4.3.4   Normal Mixtures and Conclusion

In Example 1 of *GJ*, also cited in Titterington et al. (1985), section 5.5, a mixture density of the form

$$f_\theta(x) = \frac{p}{\sigma_1}\phi\left(\frac{x-\mu_1}{\sigma_1}\right) + \frac{1-p}{\sigma_2}\phi\left(\frac{x-\mu_2}{\sigma_2}\right) \qquad (4.37)$$

is given with $\theta \equiv \mu_1$ and $\phi(x) = (1/\sqrt{2\pi})\exp(-x^2/2)$. That is, varying a single parameter in the mixture, keeping other parameters known, in fact taking values $\sigma_1 = 1, \mu_2 = 8, \sigma_2 = 4, p = 0.4$, the "true" value is chosen to be $\theta = -3$. In the simulation of *GJ*, with sample sizes $n = 5\,000$ and simulating 500 data sets, those authors begin their numerical algorithms from say an initial parameter $\theta_{o1}^{(n)} \equiv -3$ to find a "global" maximum and then start from the second initial point for their algorithm to be $\theta_{o2}^{(n)} \equiv 6$ to observe the "local" maximum, they arrive at average values of -2.993 which corresponds to the true value of -3, but their estimate of the other local root is 6.4778. There is a discrepancy between this value and the root of the asymptotic equation $K_{F_\theta}(\tau) = 0$. By reverting to numerical integration, this local root is in fact $\theta_1 = 8.0$ to one decimal place, which goes to show that the numerical algorithm of *GJ* strayed from the "local" root. See also Example 2 of *GJ*.

Many books, including Everitt and Hand 1981, Titterington et al. 1985, McLachlan and Basford 1988, and McLachlan and Peel (2000), explain the problem of multiple roots of the maximum likelihood equations. In fact the likelihood surface, say, in estimating all five parameters in Eq.(4.37) has asymptotes, meaning that the likelihood surface is unbounded. This is a principal motivation for the use of many alternative minimum distance methods for estimation of mixtures [Choi and Bulgren 1968, MacDonald 1971, Beran 1977, Clarke and Heathcote (1978,1994), Woodward et al. 1984, and Cutler and Cordiero-Braña 1996]. From a robustness point of view Clarke (2000b), pp. 474–479, demonstrates that minimum distance methods may also yield equations with multiple roots. Also for M-estimators formula (4.12) may indeed be too messy to implement. A simpler idea is to use the Cramér–von Mises statistic to choose from multiple roots. See the recommendation of Clarke and Heathcote (1994), p. 92, and also the discussion later in Section 6.1 around formulae (6.4) and (6.5).

In the conclusion of this section, we have shown the method of *GJ* is a special case in the theory of M-estimation and can be improved on even in likelihood examples by a robust choice of selection functional. The "mooted" choice of selection functional in Clarke (1990), effectively choosing *1.*, is evidence of this in Examples 4.1 and 4.2. Example 4.3 shows that *GJ*'s choice still has some merit, since *1.* fails. The current work builds on the work of *GJ* who were inspired by White (1982). It is expected the understanding of the selection functional and the link to asymptotically unique consistent roots of estimating equations will lead to further research. While the Cauchy example considered in this section is definitely in the framework of robustness theory, the example from Amemiya involves non-robust statistics. Nevertheless, it serves a purpose in that it illustrates in a relatively easy manner the theory of multiple roots when there can be several roots of one's estimating equations. Since the parametric models under consideration these days are more complex than in yesteryear, thanks to the advance of the computational power of computers, we can be assured that many examples will arise where one needs to choose a root from multiple roots of one's estimating equations.

5

DIFFERENTIABILITY AND BIAS REDUCTION

5.1 DIFFERENTIABILITY, BIAS REDUCTION, AND VARIANCE ESTIMATION

Many statistics in the literature are known to satisfy a bias expansion. For example, given a vector statistic $\hat{\theta}(X_1, \ldots, X_n) \equiv T[F_n]$, say, we know that if the statistic has been defined correctly to be Fisher consistent, then when evaluated at the whole population F_θ then $T[F_\theta] = \theta$. But in finite samples it may satisfy an expansion of the form

$$\text{bias}(\theta) = E_\theta[\hat{\theta} - \theta] = \frac{B_1(\theta)}{n} + \frac{B_2(\theta)}{n^2} + O\left(\frac{1}{n^3}\right), \qquad (5.1)$$

where E_θ represents expectation with respect to the underlying parametric population. Small sample bias terms can and do affect the performance of the statistic, and much effort is centered on negating the effects of such biases.

5.1.1 The Jackknife Bias and Variance Estimation

Quenouille (1949, 1956) introduced the idea of correcting for bias in time series estimation, and this was taken up more generally and popularized by Tukey (1958).

Robustness Theory and Application, First Edition. Brenton R. Clarke.
© 2018 John Wiley & Sons, Inc. Published 2018 by John Wiley & Sons, Inc.
Companion website: www.wiley.com/go/clarke/robustnesstheoryandapplication

Defining $\hat{\theta}_{(i)} = \hat{\theta}(X_1, \ldots, X_{i-1}, X_{i+1}, \ldots, X_n)$ to be the estimate based on the sample with the ith observation deleted, call it the ith jackknife sample, and writing

$$\hat{\theta}_{(\cdot)} = \frac{1}{n} \sum_{i=1}^{n} \hat{\theta}_{(i)},$$

we have that the jackknife estimate of bias of $\hat{\theta}$ in estimating θ is

$$\hat{b}_{\text{Jack}}(\hat{\theta}) = (n-1)[\hat{\theta}_{(\cdot)} - \hat{\theta}],$$

a bias corrected estimate of θ is

$$\hat{\theta}_{\text{corrected}} = \hat{\theta} - \hat{b}_{\text{Jack}}(\hat{\theta}) = n\hat{\theta} - (n-1)\hat{\theta}_{(\cdot)}.$$

The jackknife removes the first-order term $B_1(\theta)$ in the bias expansion of the original estimator. For example, the classical estimator for variance usually constructed as the maximum likelihood estimator (MLE) of σ^2 assuming a normal location and scale parametric family is

$$\hat{\theta} = \frac{1}{n} \sum_{i=1}^{n} (X_i - \overline{X})^2 \equiv \hat{\sigma}^2$$

reveals a jackknife bias corrected estimator

$$\hat{\theta}_{\text{corrected}} = \frac{1}{n-1} \sum_{i=1}^{n} (X_i - \overline{X})^2 \equiv s_n^2.$$

In this case the correction removes all the bias.

The jackknife approach also addresses the question of obtaining estimates of variance of estimators. Under some conditions these variance estimates are robust. Defining the ith pseudo value as

$$\tilde{\theta}_i = n\hat{\theta} - (n-1)\hat{\theta}_{(i)}$$

the average of the pseudo values is

$$\hat{\theta}_{\text{corrected}} = \frac{1}{n} \sum_{i=1}^{n} \tilde{\theta}_i = n\hat{\theta} - (n-1)\hat{\theta}_{(\cdot)},$$

which is the jackknife bias-corrected estimate. From p. 95 of Hampel et al. (1986) it is seen that the multivariate estimate

$$V_n = \frac{1}{n-1} \sum_{i=1}^{n} (\tilde{\theta}_n - \hat{\theta}_{\text{corrected}})(\tilde{\theta}_n - \hat{\theta}_{\text{corrected}})^T$$

$$\approx \text{var}_{F_n}[\text{IF}(X, F_n, T)]$$

which estimates

$$\sum(F_\theta, T) = \int \mathrm{IF}(x, F_\theta, T)\mathrm{IF}(x, F_\theta, T)^T \mathrm{d}F_\theta(x) \tag{5.2}$$

the asymptotic variance of the estimator. The estimate of variance proffered by Tukey (1958) of the actual estimator, and given in multivariate form by Efron (1982), was

$$\hat{V}_{\mathrm{Jack}}(\hat{\theta}) = \frac{n-1}{n} \sum_{i=1}^{n} (\hat{\theta}_{(i)} - \hat{\theta}_{(\cdot)})(\hat{\theta}_{(i)} - \hat{\theta}_{(\cdot)})^T$$

which in fact in terms of the pseudo values is the same as

$$\hat{V}_{\mathrm{Jack}}(\hat{\theta}) = \frac{V_n}{n}.$$

The jackknife works for both robust and non-robust statistics at the parametric model subject to some smoothness assumptions. Huber (1981, p. 16) is at pains to warn that the jackknife requires a smooth functional of F, which is not the case for quantiles. For instance, in the case of the median, the jackknife variance is unbiased for *twice* the true variance, thus failing in a spectacular manner. For estimators defined from real data, Parr (1985) and Shao and Wu (1989) provide a strong Fréchet derivative sufficient for results on the consistency of the jacknifed variance estimate. Briefly from Parr (1985), a real valued functional T is said to possess a strong Fréchet differential at F_θ if and only if there exists a $\psi : \mathcal{E} \to \mathcal{E}$ depending only on T and F_θ such that with $||G - F||$ defined to be $d_k(F, G) = ||F - G||_\infty$ we have

$$\frac{|T[G] - T[H] - \int \psi(x)\mathrm{d}[G - H](x)|}{||G - H||} \to 0 \tag{5.3}$$

as $||G - F_\theta|| + ||H - F_\theta|| \to 0$ for $G, H \in \Lambda$, where Λ is a convex set of distributions containing F_θ and all discrete distributions with finite support. According to Parr, it is not necessary to assume ψ to be bounded. For instance, we can consider the mean functional whereby $\hat{\theta} = \overline{X} = T[F_n]$, for which the jackknifed variance estimate is $\hat{V}_{\mathrm{Jack}} = s_n^2/n$. This is a classic non-robust estimator, yet it is a consistent estimator of the variance in that $n\hat{V}_{\mathrm{Jack}} \to_{a.s.} \sum(F_\theta, T) = \sigma^2$, assuming say that F_θ is a normal distribution with mean $\theta \equiv \mu$ and variance σ^2.

Parr (1985) points out that if T is Fréchet differentiable in a neighborhood of F_θ and that the derivative is continuous at F then the functional has a strong differential. In view of the development from conditions **W** that lead up to Theorem 2.5, see Clarke (2000a), it can be seen that M-estimators with bounded and continuous Ψ-functions and partial derivatives, which are also bounded and continuous, can lead to estimators that are Fréchet differentiable at each F in a small Kolmogorov neighborhood of F_θ. This then gives that such estimators will have consistent jack-knifed variance estimates, in view of existence of the strong differential, and is a valid approach to estimating the variance of robust M-estimators say. The arguments for jackknifing at the parametric model given in Shao (1989, 1993) and Shao and Tu (1995) give less stringent conditions for the consistency of the jackknife variance estimates essentially by using more generally applicable norms such as

$$||F - G||_{(\infty+l)} = ||F - G||_\infty + \left[\int |F - G|^l dx \right]^{1/l}.$$

Strong Fréchet differentiability with respect to the supremum norm automatically implies strong Fréchet differentiability with respect to $|| \cdot ||_{(\infty+l)}$.

An alternative estimator of the variance of any estimator, $T[F_n]$, is to use $\text{var}_{F_n}[\text{IF}(X, F_n, T)]$ as recommended by Huber (1981, p. 149), Hampel et al. (1986, p. 95) and Staudte and Sheather (1990, p. 79). Bednarski (1993a) stresses that if a functional is Fréchet differentiable with respect to the supremum norm or equivalently the Kolmogorov distance metric, then the asymptotic variance for the resulting estimator can be approximated via the influence function for the whole infinitesimal neighborhood. It is clear from Huber (1981), Clarke (Clarke1983a, Clarke1986b), Bednarski (1993a), and Bednarski et al. (1991) that the Fréchet differentiable functionals in this case have bounded influence functions, whereby robustness of the variance estimates therefore follows.

The actual choice of variance estimator, one based on $\hat{V}_{\text{Jack}}$ or one based on $\text{var}_{F_n}[\text{IF}(X, F_n, T)]$, may of course be predicated on whether or not one has the influence function in an explicit and analytically tractable form. For example, even with M-estimators defined through a known Ψ-function, it is necessary to have an explicit form for the matrix $M(\theta)$ in Eq. (2.35) in order to easily implement this latter estimate of variance. Such matrices are not always easily at hand.

5.1.2 Simple Location and Scale Bias Adjustments

In many often simple and explicit cases of estimators, the first or second or even more bias terms can be calculated explicitly and accounted for by making a simple subtraction. Yet with implicitly defined estimators as with those defined by Eq. (1.20) such adjustments even for the first order bias term need to be extracted from the equations. A detailed account of this for Huber's Proposal 2 estimator of Eq. (1.24) is given in Clarke and Milne (2004). For example, the MLE of location and scale of the normal distribution is given by $\Psi(x) = (x, -1 + x^2)^T$, for which it is known that the subsequent estimate of variance $\hat{\sigma}^2$ has a bias of $-\sigma^2/n$. Perhaps surprisingly to some, the bias corrected variance estimate s_n^2, while being unbiased for σ^2 is not however such that the corresponding scale estimate, s_n, is unbiased for σ. In fact, since $\text{var}[s_n] = E[s_n^2] - E[s_n]^2 = \sigma^2 - E[s_n]^2 > 0$, it is the case that

$E[s_n] < \sigma$. Along the lines of Problem 5.3, it is not difficult to show that for the MLE for scale, $\hat{\sigma} = (\hat{\sigma}^2)^{\frac{1}{2}}$, the exact first order bias term is $-\frac{3}{4}\sigma/n$. Yet this bias can be calculated in another way, by making appropriate expansions, which are now illustrated and have further use.

In the case for M-estimators of scale, when location and scale are estimated simultaneously, it is a little more complex to work out the expectations, due to the estimators only being implicitly defined through Eq. (2.1). Clarke and Milne (2004), following the lead of Cox and Hinkley (1974, pp. 260–309) in a similar problem, resolve to make Taylor expansions of the component equations in (2.1) evaluated at $T[F_n] = (\hat{\mu}, \hat{\sigma})^T$ and, defining $\mathbf{K}_{F_n}(T[F_n]) = (K_{n1}(\hat{\mu}, \hat{\sigma}), K_{n2}(\hat{\mu}, \hat{\sigma}))^T$, expand the second component equation for scale around the true parameters $\theta^T = (\mu, \sigma)$. The notation $\partial^2_{\mu\mu}$ is used to denote the second partial derivative with respect to μ, and similarly $\partial^2_{\sigma\sigma}$ to denote the second partial derivative with respect to σ. The second partial derivative with respect to both μ and σ is then denoted $\partial^2_{\mu\sigma}$. One then has the expansion

$$0 = K_{n2}(\hat{\mu}, \hat{\sigma}) = K_{n2}(\mu, \sigma) + (\hat{\mu} - \mu)\partial_\mu K_{n2}(\mu, \sigma) + (\hat{\sigma} - \sigma)\partial_\sigma K_{n2}(\mu, \sigma)$$

$$+ \frac{1}{2}\{(\hat{\mu} - \mu)^2\partial^2_{\mu\mu}K_{n2}(\mu, \sigma) + 2(\hat{\mu} - \mu)(\hat{\sigma} - \sigma)\partial^2_{\mu\sigma}K_{n2}(\mu, \sigma)$$

$$(\hat{\sigma} - \sigma)^2\partial^2_{\sigma\sigma}K_{n2}(\mu, \sigma)\} + O_p(n^{-\frac{3}{2}}). \tag{5.4}$$

Taking expectation of this equation and also a similar equation involving K_{n1} yields an explicit formula for $E[\hat{\theta} - \theta]$. For instance, use is made of the approximation

$$n^{\frac{1}{2}}\begin{bmatrix} \hat{\mu} - \mu \\ \hat{\sigma} - \sigma \end{bmatrix} \approx -M(\mu, \sigma)^{-1}\frac{1}{n^{\frac{1}{2}}}\begin{bmatrix} \sum_{i=1}^{n}\psi_1(\frac{X_i-\mu}{\sigma}) \\ \sum_{i=1}^{n}\psi_2(\frac{X_i-\mu}{\sigma}) \end{bmatrix} \tag{5.5}$$

where

$$M(\mu, \sigma) = \begin{bmatrix} \int\{\partial_\mu\psi_1(\frac{x-\mu}{\sigma})\}dF_{\mu,\sigma}(x) & \int\{\partial_\sigma\psi_1(\frac{x-\mu}{\sigma})\}dF_{\mu,\sigma}(x) \\ \int\{\partial_\mu\psi_2(\frac{x-\mu}{\sigma})\}dF_{\mu,\sigma}(x) & \int\{\partial_\sigma\psi_2(\frac{x-\mu}{\sigma})\}dF_{\mu,\sigma}(x) \end{bmatrix}$$

The expansion (5.5) is justified for $\Psi = (\psi_1, \psi_2)^T$ defining Huber's Proposal-2 or the three-part redescender of Eqs. (1.21) and (1.22) through the arguments in Clarke (1986b) illustrating Fréchet differentiability of these estimators, for instance at the normal distribution. See Example 3.1. Then use is made of the Fréchet expansion.

Taking the expectation of the Taylor series expansion for the second component equation in Eq. (1.20) involving $K_{n2}(\hat{\mu}, \hat{\sigma})$ and making use of the approximation (5.5), details in Clarke and Milne (2004) lead to the following formula for the first-order bias term of the M-estimator of scale,

$$
E_{\mu,\sigma}(\hat{\sigma} - \sigma) = \frac{\sigma}{n} \frac{1}{E(Z\psi_2')} \left(-\frac{E(\psi_2'\psi_1)}{E(\psi_1')} - \frac{E(Z\psi_2'\psi_2)}{E(Z\psi_2')} \right.
$$
$$
\left. + \frac{E(\psi_1^2)E(\psi_2'')}{2E(\psi_1')^2} + \frac{E(\psi_2^2)(2E(Z\psi_2') + E(Z^2\psi_2''))}{2E(Z\psi_2')^2} \right)
$$
$$
+ O(n^{-2}). \tag{5.6}
$$

Here Z is the standardized variable and E represents expectation with respect to the standardized distribution, in this case for example $E(Z\psi_2') = \int x\psi_2'(x)d\Phi(x)$. Also $E_{\mu,\sigma}$ is the expectation with respect to the unstandardized distribution.

Remark 5.1: The above formula remains valid if $\Phi(x)$ is replaced by other suitably regular symmetric distributions $F(x)$.

Remark 5.2: Equation (5.6) must be interpreted at least heuristically for ψ-functions which do not have continuous derivatives, as in the case of Huber's Proposal 2 with a finite tuning constant k. Such functions are at least continuous and piecewise continuously differentiable. For example calculations we refer to Huber (1964, p. 78) and Hampel et al. (1986, p. 103). For instance $E(\psi_2'')$ and $E(Z^2\psi_2'')$ need to be interpreted this way in formula (5.6).

Formula (5.6) when evaluated at the MLE for the normal location and scale parametric family with $\Psi(x) = (x, -1 + x^2)^T$ gives the aforesaid bias term in the scale of $-\frac{3}{4}\sigma/n$. Table 5.1 taken from Clarke and Milne (2004, table 1) illustrates the bias for Huber's Proposal 2 more generally. If $\hat{\sigma}$ is the scale solution to Eq. (1.20), it follows that a bias corrected estimator of scale is $\hat{\sigma}^* = (n/(n + b^*))\hat{\sigma}$, where b^* denotes the bias parameter, for example, calculated in Table 5.1. Further values for scale bias for the three-part redescender are given in Clarke and Milne (2013).

A possible use for the bias correction of the scale M-estimator suggested in Clarke and Milne (2004) is in the use of confidence intervals for location based on the asymptotic normal distribution when jointly estimating location and scale. From the expansion (5.5) the asymptotic distribution leads to a 95% confidence interval for location of

$$
\left(\hat{\mu} - 1.96\frac{\hat{\sigma}}{\sqrt{n}}\lambda, \ \hat{\mu} + 1.96\frac{\hat{\sigma}}{\sqrt{n}}\lambda \right), \quad \text{where } \lambda^2 = \frac{E(\psi_1^2)}{E(\psi_1')^2}.
$$

TABLE 5.1   First-Order Bias Term, with Associated Asymptotic Variances of Location and Scale, for Huber's Proposal 2 at the Standard Normal Distribution

k	b^* = First-Order Bias Parameter for Scale	Asymptotic Variance of Location	Asymptotic Variance of Scale
1.000	−0.7424	1.1073	1.0335
1.285	−0.6669	1.0600	0.7944
1.400	−0.6512	1.0466	0.7326
1.500	−0.6423	1.0371	0.6894
1.645	−0.6357	1.0262	0.6402
1.750	−0.6347	1.0202	0.6123
1.960	−0.6397	1.0116	0.5710
3.000	−0.7137	1.0004	0.5045
$+\infty$	−0.7500	1.0000	0.5000

Source: Clarke and Milne (2004). Reproduced with the permission of John Wiley & Sons.

The confidence interval involves the estimate of σ. The use of the bias-corrected scale estimator $\hat{\sigma}^*$ leads to confidence levels closer to nominal values. However, such results are based on the confidence interval constructed from the asymptotic distribution rather than the Student's-distribution. It is an appropriate time to now introduce a technique that has revolutionized variance and confidence interval calculations.

5.1.3   The Bootstrap

The bootstrap has close connections to the jackknife and was introduced in Efron (1979). Bootstrapping is a resampling technique that takes more than one form, but essentially it is one of two approaches in order to obtain an estimate of bias and of variance. The theory also extends to confidence intervals. The main approaches are the parametric bootstrap and the nonparametric bootstrap. There is not sufficient space in this book to explain all the facets of this methodology. Books by Efron and Tibshirani (1993), Davison and Hinkley (1997), Shao and Tu (1995), and a technical work by Hall (1992) explain where the subject began and its wide application.

We shall describe the bootstrap for a univariate parameter though there is no essential difficulty in extending the bootstrap to more complicated models. The rudiments of the bootstrap are as follows. Assume the underlying distribution is F and the bias of an estimator is

$$\text{bias}(\hat{\theta}) = E_F[\hat{\theta}] - \theta = E_F[T[F_n]] - T[F],$$

which is a functional of the underlying distribution and also the empirical distribution function. One applies a "plug-in" principal involving the replacement of population quantities by sample quantities and sample quantities by resample quantities. Consider the following algorithm also described for the special case of the exponential distribution in Staudte and Sheather (1990, p. 34). We make the natural adaption to a more general parametric estimator other than $\overline{X}$ and assume a more general parametric distribution.

1. For each $b = 1, \ldots, B$, generate n independent observations from $F_{\hat{\theta}}$, where $\hat{\theta} = T[F_n]$ is the estimate based on the original data, and form a bootstrap estimate $\hat{\theta}_b$.
2. Find the mean and variance of the bootstrap estimates:

$$\overline{\theta}_B = \frac{1}{B} \sum_{b=1}^{n} \hat{\theta}_b$$

$$s_B^2 = \frac{1}{B-1} \sum_{b=1}^{B} (\hat{\theta}_b - \overline{\theta}_B)^2.$$

Then the parametric bootstrap estimate of bias is

$$\hat{b}_{\text{Boot}}(\hat{\theta}) = \overline{\theta}_B - \hat{\theta},$$

and the bootstrap bias-corrected estimate is

$$\hat{\theta}_{\text{Boot,corrected}} = \hat{\theta} - \hat{b}_{\text{Boot}}(\hat{\theta}) = 2\hat{\theta} - \overline{\theta}_B,$$

while the parametric bootstrap estimate of standard error is

$$\hat{\text{SE}}_{\text{boot}}[T[F_n]] = s_B.$$

Thus, the parametric bootstrap estimate is simply s_B, the standard deviation of the B estimates obtained by sampling from the parametric distribution evaluated at or using the original estimate of the parameter.

We further quote with permission Staudte and Sheather (1990, p. 35) discussion of "The Nonparametric Bootstrap." This "is obtained in the same way as the parametric bootstrap in 1, except that sampling is from the empirical distribution function F_n." Then "Thus no knowledge of the underlying parametric form of F is required. This method is most frequently employed and is often simply referred to as 'the bootstrap'. The bootstrap estimate depends on B and the random number generator. For bias and variance estimation, around 25–50 bootstrap replications are usually sufficient, though the more the better and usually as always the prescription will tend to depend on the type of estimator. Maronna et al. (2006, p. 141) warn of the impractical nature of the bootstrap for many robust statistics that require large computing times, and there are occasions when the bootstrap sample might contain a higher proportion of outliers than the original sample which leads "to quite incorrect values of the recomputed estimate." See also the discussion of Salibian-Barrera and Zamar (2002) with suggestions to avoid this in regression.

5.1.4   The Choice to Jackknife or Bootstrap

Historically, the jackknife is a precursor to the bootstrap. Well-known functionals that fail for the jackknife are the median and quantiles which incidentally have discontinuous influence functions. The bootstrap is known to work for such functionals. The bootstrap is consistent under weaker conditions than Fréchet differentiability, and in fact compact differentiability-type conditions appear to be sufficient as indicated by Yang (1985), Lohse (1987), Gill et al. (1989), and Shao and Tu (1995). This includes the mean functional but also seemingly robust functionals such as the median and quantiles. The application is therefore quite broad, including both non-robust and robust functionals.

The mean functional is not a robust functional. It satisfies a linear expansion exactly but is not a continuous functional. Nevertheless, it has a strong Fréchet expansion in the sense of Parr (1985), for this and other functionals with unbounded influence functions can satisfy strong Fréchet expansions. If they do, this is usually done by restricting the domain set Λ on which they are defined. Jackknife consistency is then given by Parr (1985, Theorem 1). Of course there is still a question of whether such functionals are useful, since in terms of possible asymptotic bias, at least of M-estimators, the gross error sensitivity $\gamma^* = +\infty$ indicates such estimators can be very sensitive, and especially so if it is known that small departures from the parametric model are possible.

Since the jackknife requires fewer computations, it may be preferable in situations where both can be applied. This is certainly true in terms of iteratively derived estimates, such as for implicitly defined functionals, which many robust estimators are. The situation does become a little complex when there can be multiple minima of one's objective function, or equivalently, multiple solutions of one's estimating equations. Potentially, both the jackknife and the bootstrap are able to stray from a solution of interest, since one's algorithm is resampling. One or two resultant resample estimates can potentially diverge a long way away from the rest, and when taking the average of all of them on the way to getting one's jackknife or bootstrap estimate, the average being a non-robust estimator will not reflect the true properties of the original estimator.

In a way this is to be expected. For complicated models in the presence of multiple solutions, a good applied statistician would start from a number of initial estimates corresponding to regions on the parameter space which reflect physical phenomena that can be interpreted by the parametric model, before iterating to hone in on what may be the most appropriate solution for one's estimator. This process could be arduous if we were to repeat it for each jackknife resample or each bootstrap resample for instance. But what can be done about it? There may be hope in terms of jackknifed samples from robust M-functionals when starting one's jackknife estimates from the same robust initial estimate. To illustrate this we refer to results for M-estimators, but such results could be equally applied to other estimators including those involving density estimates, anywhere where the theory of uniform convergence applies. The Newton algorithm, also known as the Newton–Raphson algorithm in the univariate parameter setting, is a well-known numerical algorithm used to find stationary points of objective functions, or solutions of one's estimating equations. Studying uniform convergence, as has been done in earlier chapters, can show that the algorithm is bound to converge to the asymptotically unique consistent root of the equations starting from within the same small ball about the solution to the asymptotic equations. Hence, since our initial estimate being an estimate or solution obtained from the full sample should be in that small ball for all sufficiently large n by the consistency results of Chapter 2, then for each set of equations from each jackknifed sample by uniform convergence the iteration should converge to the local asymptotically unique root of the estimating equations corresponding to the jackknifed estimate. For the result to be robust, we must use psi-functions that are bounded and continuous, and to facilitate the argumentation we should have continuous and bounded partial derivatives.

5.2   FURTHER RESULTS ON THE NEWTON ALGORITHM

Following the spirit of Clarke (1986a) which thoroughly investigated the domain of attraction of the Newton–Raphson algorithm applied to the estimating equations associated with various M-estimators of location, Clarke and Futschik (2007) studied a set of general equations. For example, suppose that we want to estimate a parameter τ_0 defined as the solution of the equations $\lambda(\tau) = 0$, where $\lambda : D \subseteq \mathcal{E}^r \to \mathcal{E}^r$. We assume the solution to be unique in some open ball B_0 of interest, but there may exist further solutions outside of B_0. Assume furthermore that a uniformly consistent estimate λ_n of λ is available at least on $\overline{B}_0$, the closure of B_0. Then a natural estimate of τ_0 is $\hat{\tau}_{0n}$ where $\hat{\tau}_{0n}$ satisfies $\lambda_n(\hat{\tau}_{0n}) = 0$. Again $\hat{\tau}_{0n}$ is not necessarily unique. Frequently however, in such a setting, there is an asymptotically unique consistent root of the equations $\lambda_n(\tau) = 0$. By this we mean that there is some ball B_δ of radius $\delta > 0$ about τ_0 where $\hat{\tau}_{0n}$ exists and is unique on B_δ for all sufficiently large n, and furthermore that $\hat{\tau}_{0n}$ converges to τ_0 almost surely. The $\hat{\tau}_{0n}$ solving

$$\lambda_n(\tau) = 0 \tag{5.7}$$

may be hard or even impossible to obtain directly in higher dimensions. A common approach is thus to use Newton's algorithm starting from some initial estimate y_0. For solving $\lambda_n(\tau) = 0$, the algorithm is defined by the iteration

$$y_{l+1}^{(n)} = y_l^{(n)} - [\lambda_n'(y_l^{(n)})]^{-1}\lambda_n(y_l^{(n)}). \quad l = 0, 1, 2, \dots \tag{5.8}$$

If the iteration converges, then the limit as $l \to \infty$ is taken as the estimate.

Since Newton's algorithm may in general either not converge at all or converge to another than the desired zero, it is possible to provide conditions under which a.s. convergence to $\hat{\tau}_{0n}$ holds when starting from some consistent initial estimate at least for large enough n. The reason why one may wish to use $\hat{\tau}_{0n}$ rather than say an initial consistent estimate is that the estimate $\hat{\tau}_{0n}$ may be more efficient (many examples of this are found in the literature). Indeed under suitable choices of λ_n, a one step or even several step Newton iteration is frequently quoted as retaining the efficiency of the estimator based on Eq. (5.7).

It is shown in Clarke and Futschik (2007) under easily verifiable conditions and illustrated in many examples that there is a fixed neighborhood based on the asymptotic curve $\lambda(\tau)$ in which for all sufficiently large n the Newton iteration would remain and indeed if the iteration was allowed to continue until fully iterated then one is assured that one will converge to the unique consistent root.

Restricting our attention to M-estimators discussed in section 3.1 of Clarke and Futschik (2007), and more importantly to robust M-estimators with bounded Ψ-functions, we have the following result for jackknifed estimates obtained from the Newton algorithm.

Theorem 5.1: *(Jackknife consistency of Newton iteration) Consider the M-estimator for which the defining Ψ-function satisfies **Conditions W** of section 2. Then beginning from the asymptotically unique consistent root, $\hat{\theta} = T[F_n]$, and denoting $\hat{\theta}_{(i)} = \hat{\theta}(X_1, \dots, X_{i-1}, X_{i+1}, \dots, X_n)$ to be the estimate that is obtained from the limit of the Newton algorithm applied to the M-estimating equations from the ith jackknifed sample beginning from the initial estimate $\hat{\theta}$. Then there exists a non-empty open ball set $B_\delta(\theta)$ such that for all sufficiently large n the statement*

$$\sup_{1 \le i \le n} |\hat{\theta}_{(i)} - \theta| < \delta$$

holds, and the jackknifed variance estimate is consistent.

Proof. The M-estimating equations for the ith jackknife sample are

$$K_n^{(i)}(\tau) = \frac{1}{n-1} \sum_{j \neq i} \Psi(X_j, \tau) = 0.$$

Since

$$K_n^{(i)}(\tau) = \frac{nK_{F_n}(\tau) - \Psi(X_i, \tau)}{n-1}$$

$$= K_{F_n}(\tau) + \frac{K_{F_n}(\tau) - \Psi(X_i, \tau)}{n-1},$$

we deduce

$$\sup_{1\leq i\leq n}\sup_{\{\tau\in\Theta\}} |K_n^{(i)}(\tau) - K_{F_n}(\tau)| \leq \frac{2}{n-1} \sup_{\{(x,\tau)|x\in R, \tau\in\Theta\}} |\Psi(x,\tau,)| \to 0.$$

Similarly, we deduce

$$\sup_{1\leq i\leq n}\sup_{\{\tau\in\Theta\}} |\frac{\partial}{\partial\tau}K_n^{(i)}(\tau) - \frac{\partial}{\partial\tau}K_{F_n}(\tau)| \to 0.$$

Thus, for example, we have by the triangle inequality that

$$\sup_{\{\tau\in\Theta\}}\sup_{1\leq i\leq n} |K_n^{(i)}(\tau) - K_G(\tau)| \leq \sup_{\{\tau\in\Theta\}} |K_{F_n}(\tau) - K_G(\tau)|$$

$$+ \sup_{\{\tau\in\Theta\}}\sup_{1\leq i\leq n} |K_n^{(i)}(\tau) - K_{F_n}(\tau)|.$$

From this and a similar result for the partial derivative we deduce

$$\sup_{\{\tau\in\Theta\}}\sup_{1\leq i\leq n} |K_n^{(i)}(\tau) - K_G(\tau)| \to_{a.s.} 0 \tag{5.9}$$

$$\sup_{\{\tau\in\Theta\}}\sup_{1\leq i\leq n} |\frac{\partial}{\partial\tau}K_n^{(i)}(\tau) - \frac{\partial}{\partial\tau}K_G(\tau)| \to_{a.s.} 0. \tag{5.10}$$

The uniform approximation of the estimating curves $K_n^{(i)}(\tau)$ to the same asymptotic curve $K_G(\tau)$, and similarly for their partial derivatives, implies that conditions of Theorem 1 of Clarke and Futschik (2007) are met uniformly over jackknifed samples. The argumentation goes the same way as in that paper. As has been noted from the discussion of Section 5.1.1, the jackknifed estimates are consistent since they are an asymptotically unique consistent root in a small ball $B_\delta(\theta)$. Example application of this theory on real data is in Clarke et al. (2017, Example 2). To understand that application, we now embark in the next chapter on a discussion of minimum distance estimators.

PROBLEMS

5.1. Consider the made up sample of size $n = 3$.

$$\{X_1 = 1, X_2 = 2, X_3 = 3\}$$

Show for this data the biased estimate of variance $\hat{\sigma}^2 = \frac{2}{3}$ and the unbiased estimate of variance $s_n^2 = 1$ and letting $\hat{\theta} \equiv \hat{\sigma}^2$ show $\hat{\theta}_{(1)} = \frac{1}{4}$, $\hat{\theta}_{(2)} = 1$,

$\hat{\theta}_{(3)} = \frac{1}{4}$ and $\hat{\theta}_{(\cdot)} = \frac{1}{2}$. Hence, show $\hat{b}_{\text{Jack}}(\hat{\theta}) = -\frac{1}{3}$ and so $\hat{\theta}_{\text{corrected}} = \hat{\theta} - \hat{b}_{\text{Jack}}(\hat{\theta}) = 1$. Also show the pseudo values are $\tilde{\theta}_1 = \frac{3}{2}$, $\tilde{\theta}_2 = 0$ and $\tilde{\theta}_3 = \frac{3}{2}$. Hence, see also that $\hat{\theta}_{\text{corrected}} = \frac{1}{3}(\tilde{\theta}_1 + \tilde{\theta}_2 + \tilde{\theta}_3) = 1$.

5.2. Show that for the estimator $\hat{\theta} = \overline{X}$ that $\hat{V}_{\text{Jack}} = \frac{s_n^2}{n}$.

5.3. Consider $X_1, \ldots, X_n$ to be an *i.i.d.* sample with common distribution from the normal location and scale family

$$\mathcal{F} = \left\{ \frac{1}{\sigma}\phi\left(\frac{x-\mu}{\sigma}\right) \mid -\infty < \mu < \infty, \ 0 < \sigma < \infty \right\}.$$

A description of the following parts of problems are found in the introduction to Clarke and Milne (2004).

(a) Derive the MLE for $\theta = (\mu, \sigma)$ and show

$$\hat{\mu} = \overline{X} \text{ and } \hat{\sigma} = \sqrt{\frac{1}{n}\sum_{i=1}^{n}(X_i - \overline{X})^2}.$$

(b) Show $\hat{\mu}$ is unbiased for μ and that $\hat{\sigma}^2$ is biased for σ^2 with bias $-\sigma^2/n$.

(c) Show the following relation for the mean squared error of $\hat{\sigma}^2$:

$$\text{MSE}(\hat{\sigma}^2) = E[(\hat{\sigma}^2 - \sigma^2)^2] = \frac{(2n-1)}{n^2}\sigma^4 = \frac{2\sigma^4}{n} - \frac{\sigma^4}{n^2}.$$

(d) Consider the Taylor expansion

$$f(\hat{\sigma}^2) = f(\sigma^2) + f'(\sigma^2)(\hat{\sigma}^2 - \sigma^2) + \frac{1}{2}f''(\sigma^2)(\hat{\sigma}^2 - \sigma)^2 + O((\hat{\sigma}^2 - \sigma^2)^3)$$

and substitute in the function $f(x) = \sqrt{x}$ and take expectations of both sides to show that

$$E[\hat{\sigma}] \approx \sigma + \frac{1}{2\sigma}\left(-\frac{\sigma^2}{n}\right) - \frac{1}{8\sigma^3}E[(\hat{\sigma}^2 - \sigma^2)^2].$$

(e) Using the above calculations and ignoring terms of order $1/n^2$ establish that

$$E[\hat{\sigma} - \sigma] \approx -\frac{3\sigma}{4n}.$$

6

MINIMUM DISTANCE ESTIMATION AND MIXTURE ESTIMATION

6.1 MINIMUM DISTANCE ESTIMATION AND REVISITING MIXTURE MODELING

While the theory of robust statistics had been developing with the appearance of the works by Tukey (1960), Huber (1964), and Hampel (1968), there had been also an alternative line of development of strongly consistent estimators at the model distribution some time earlier by proponents of minimum distance estimation methods. This was even prior to the general advent of the computer. Works by Wolfowitz (1953, 1957), Matusita (1953), Kac et al. (1955), and Blackman (1955) opened up a new line of research on estimators, other than the maximum likelihood estimator, that converge almost surely to the true parameter. While the theory of such estimators received some further attention in say Sahler (1970), Bölthausen (1977), and Pollard (1980), it was not until works by Beran (1977), Millar (1981), Parr and Schucany (1980), Parr and DeWet (1981), and Boos (1981) that the theory of minimum distance estimation and the theory of robust estimation can be said to have truly began to overlap properly. Recognition that estimation in the parametric framework was possible for some quite complicated parametric models has already been noted in Section 4.3.4 in the works of Choi and Bulgren (1968) and Macdonald (1971) for mixture modeling. These works had their own peculiar

Robustness Theory and Application, First Edition. Brenton R. Clarke.
© 2018 John Wiley & Sons, Inc. Published 2018 by John Wiley & Sons, Inc.
Companion website: www.wiley.com/go/clarke/robustnesstheoryandapplication

reason for being written due to the singularity problems with the maximum like-lihood estimator. Under certain conditions for mixture models Choi and Bulgren (1968) describe the asymptotic theory of the statistic that minimizes the distance measured for real-valued data as

$$\int_{-\infty}^{+\infty} \{F_n(x) - F_\tau(x)\}^2 dF_n(x). \tag{6.1}$$

Their interpretation of the integration yields a distance that minimizes

$$\frac{1}{n} \sum_{i=1}^{n} \{\frac{i}{n} - F_\tau(X_{(i)})\}^2. \tag{6.2}$$

The importance of the contribution of Macdonald (1971) is that it is recognized with the appropriate interpretation of the integration that minimizing the statistic Eq. (6.1) is in fact the same as minimizing the Cramér–von Mises distance statistic, since Eq. (6.1) can be interpreted as

$$\frac{1}{n} \sum_{i=1}^{n} \left\{ \frac{i - 1/2}{n} - F_\tau(X_{(i)}) \right\}^2. \tag{6.3}$$

Since the Cramér–von Mises distance is well known to be

$$\omega^2(\tau) = \int_{-\infty}^{\infty} \{F_n(x) - F_\tau(x)\}^2 dF_\tau(x) \tag{6.4}$$

$$= \frac{1}{12n^2} + \frac{1}{n} \sum_{i=1}^{n} \left\{ \frac{i - 1/2}{n} - F_\tau(X_{(i)}) \right\}^2 \tag{6.5}$$

minimizing the statistic Eq. (6.1) is the same as minimizing Eq. (6.4). The impor-tance of this is twofold. Firstly, as argued by Macdonald the estimator that min-imizes Eq. (6.4), call it MCVM, appears to have smaller small sample bias than the estimator that minimizes Eq. (6.2) for proportion estimation as demonstrated by simulation. But perhaps more importantly the Cramér–von Mises distance is well recognized in the area of goodness of fit where it was introduced. See Cramér (1928, 1946) and von Mises (1928). Indeed at the time of the initial development of the minimum distance estimators, Darling (1955) broached the subject of mini-mizing the Cramér–von Mises distance in carrying out hypothesis testing in order to gain an estimator of τ but did not advocate its use for estimation, for the reason that the maximum likelihood estimator was more efficient. The subject of good-ness of fit testing using the Cramér–von Mises statistic was well summarized in the book by D'Agostino and Stephens (1986), for example, albeit the subject has received more attention recently. It is not until the rise of the problems discussing asymptotes in the likelihood for mixtures that the Cramér–von Mises distance was seriously considered as an alternative estimator.

It was in the context of mixtures that another well cited alternative to the mixture estimation problem was elucidated. Quandt and Ramsey (1978) introduced the estimation of parameters in the model (4.37) by minimizing a distance based around a moment generating function. Adopting the notation of our book, the estimator of Quandt and Ramsey is to choose a parameter τ to minimize

$$S_n(\tau) = \int [\bar{y}_n(t) - G(t, \tau)]^2 dK_{QR}(t), \qquad (6.6)$$

where $G(t, \tau) = E[e^{tX}]$ is the moment generating function of the random variable X that has the mixture distribution Eq. (4.37) for some small in magnitude t and $\bar{y}_n(t) = \frac{1}{n} \sum_1^n \{e^{tX_j}\} = \int e^{tx} dF_n(x)$ is the empirical moment generating function. Finally, $K_{QR}(t)$ is chosen to be a weight measure that attributes positive mass to a set of points $\{t_1, \ldots, t_k\}$ found in an open interval $(a, b), a < 0 < b$. The reason why we may wish to choose such a distance is that there is a one-to-one correspondence between the moment generating function and the parametric distribution. Hopefully, by minimizing the corresponding distance between the empirical and the expected moment generating functions we will correctly estimate the parameters of the corresponding distribution function, and indeed Quandt and Ramsey go on to give the results for consistency and asymptotic normality of the resulting estimators which is supported with some simulations.

However, another estimator of the same ilk is that based around the characteristic function $\tilde{\varphi}(t, \tau) = E[e^{itX}]$. Here t is a real number and as usual E denotes expectation, while $i = \sqrt{-1}$ is the imaginary number. The one-to-one correspondence between the characteristic function and the distribution is explored even in classic texts such as Feller (1966) and Lukacs (1970). Using the characteristic function has the potential to expand the class of parametric families that can be considered. For instance the class of stable laws, which have as a class

$$\ln \tilde{\varphi}(t, \tau) = \ln E[e^{itX}]$$

$$= i\delta t - \gamma |t|^\alpha [1 + \beta(\frac{t}{|t|}) w_S(t, \alpha)], \qquad (6.7)$$

where X has a stable distribution $\tilde{S}$. The parameter space involves $\tau = (\alpha, \beta, \gamma, \delta)$, where

$$\{0 < \alpha \leq 2, \ |\beta| \leq 1, \ \gamma > 0, \ -\infty < \delta < \infty\}$$

and

$$w_S(t, \alpha) = \tan(\frac{\pi\alpha}{2}) \ \alpha \neq 1;$$

$$= \frac{2}{\pi} \ln \{|t|\}, \ \alpha = 1.$$

Here the parameter α is the characteristic exponent of the distribution, β is the skewness parameter, γ is the scale parameter, and δ is the location parameter. Any probability distribution is given by the Fourier transform of its characteristic function $\tilde{\varphi}(t)$ by

$$f(x) = \frac{1}{2\pi} \int_{-\infty}^{\infty} \tilde{\varphi}(t)\, e^{-ixt} dt. \qquad (6.8)$$

There is no general analytic solution for the form of $f(x)$. On the other hand, there are three special cases which can be expressed in terms of well-known distributions. For specific choices of parameters τ, see Voit (2005), one can arrive at the normal distribution, the Cauchy distribution, or what is known as the Lévy distribution, the latter of which is given by the formula

$$f(x, \gamma, \delta) = \sqrt{\frac{\gamma}{2\pi}} \frac{1}{(x-\delta)^{3/2}} \exp\left\{ -\frac{\gamma}{2(x-\delta)} \right\}; \ \delta < x < \infty. \qquad (6.9)$$

Thus, the stable laws offer the potential to fit a variety of distributions, including heavy tailed distributions also including skewness, and with the advent of current computing power they have a wide variety of applications. For the wider family of the stable laws, the lack of a general formula other than for specific distributions hampered the initial development of the stable law estimation theory for some time, for clearly without a closed form expression for the density in terms of the observation variable and the parameters it is not possible to carry out maximum likelihood estimation in the usual sense. But in a flurry of developments in the late 1970's, works such as Paulson et al. (1975), Leitch and Paulson (1975), Thornton and Paulson (1977), and Heathcote (1977) explore the estimation of parameters using the minimum distance approach based on distances that are to do with the characteristic function. Essentially, the proposed estimators minimized a distance based on an integrated squared error

$$I_n(\tau) = \int |\tilde{\varphi}_n(t) - \tilde{\varphi}(t, \tau)|^2 d\tilde{K}(t), \qquad (6.10)$$

where here $\tilde{\varphi}_n(t) = \frac{1}{n} \sum_{j=1}^{n} e^{itX_j}$ is the empirical characteristic function that estimates the true characteristic function of the data. The weight function chosen by Leitch and Paulson (1975) to investigate stock price data was $d\tilde{K}(t) = e^{-t^2} dt$ for example, for the reason that it forces convergence of the integral and allows Hermitian quadrature. Clarke and Heathcote (1978) note the robustness of the characteristic function based method compared to that of the moment generating function method proposed by Quandt and Ramsey, since in taking the approach based on characteristic-based functions outliers have bounded influence on the distance, whereas the influence of outliers can have an exponential influence on the distance based on the moment generating function. Moreover, it is illustrated in that comment that using a weight function with an atom of weight at one single point that the characteristic function based method can have higher efficiency for a broad part of the parameter space. Either way both methods have efficiency that is dependent on the model parameter.

In hypothesis testing, relationships between the test statistics based on characteristic functions where one chooses a weight function in Eq. (6.10) with an atom or atoms of weight at one or two points on the real line and say the Cramér–von Mises test statistic are explored, for example, in Feigin and Heathcote (1976) using some representations due to Durbin and Knott (1972). On the other hand, in estimation we may be more interested, in view of the general development of earlier chapters, in exploring relationships between the integrated squared error estimation of parameters and say M-estimators. Assuming interchange of integration and differentiation, the minimizing equations of (6.10) correspond to the choice

$$
\Psi(x, \tau) = \int [\{\cos(tx) - u(t, \tau)\} \left(\frac{\partial}{\partial \tau} \right) u(t, \tau)
$$

$$
+ \{\sin(tx) - v(t, \tau)\} \left(\frac{\partial}{\partial \tau} \right) v(t, \tau)] d\tilde{K}(t),
$$

(6.11)

where $u(t, \tau)$ and $v(t, \tau)$ are the real and imaginary parts of the characteristic function of the distribution F_τ. This can be observed from the equations of Heathcote (1977, p. 257). Such estimators when implemented with a normal parametric family and the weight function $d\tilde{K}(t) = e^{-t^2} dt$ as employed in the work of Paulson et al. (1975) and Leitch and Paulson (1975) and others give redescending psi functions, with some loss of efficiency. It is rather more the intent that one can use this methodology for estimating parameters in some heavy tailed and indeed skewed and heavy tailed parametric distributions when other methods are not easily brought to bear.

The importance of the estimation methods of minimum distance estimation that are related to M-estimators is that one has ready made ψ-functions for some quite complicated parametric models, that can lead to relatively efficient M-estimators, for which one knows one can readily obtain a solution to the M-estimating equations, since one is minimizing a distance. This has of course to be related to the parametric model at hand. One estimator, that is a variant of the Cramér–von Mises distance estimator and which has been studied for this purpose in Blackman (1955) and Knüsel (1969), is the L_2-minimum distance estimator that minimizes

$$
\int \{F_n(x) - F_\tau(x)\}^2 dK^*(x)
$$

(6.12)

for some suitable weighting measure $K^*(x)$. On the real line, and for suitable weight functions K^*, the estimator can be seen to be a solution of equations,

$$
\int -2\{F_n(x) - F_\tau(x)\} \{ \left(\frac{\partial}{\partial \tau} \right) F_\tau(x)\} dK^*(x),
$$

which on integrating by parts can be seen to be an M-estimator, since it is a solution of Eq. (2.1) where

$$\Psi(x, \tau) = \int_{\infty}^{x} \{ \left(\frac{\partial}{\partial \tau} \right) F_{\tau}(y) \} dK^{*}(y)$$

$$- \int_{-\infty}^{\infty} \int_{\infty}^{u} \{ \left(\frac{\partial}{\partial \tau} \right) F_{\tau}(y) \} dK^{*}(y) dF_{\tau}(u). \tag{6.13}$$

Blackman (1955) uses Eq. (6.12) with a weight function of Lebesgue measure, so that $dK^{*}(x) = dx$, which gives an estimator of location at the normal parametric model that has an asymptotic efficiency of approximately 96%. An example illustration of the closeness of the cumulative empirical distribution to the generating normal model located at θ is given in Figure 6.1. The estimator at the normal location model is an M-estimator with $\Psi(x) = \Phi(x) - \frac{1}{2}$ and the bounded nature of this psi function in comparison to the psi function for the MLE is illustrated in Figure 6.2. The estimator is formed from a smooth bounded ψ-function deeming by Conditions **A** that the estimator is weakly continuous and Fréchet differentiable and has all the hallmarks of a robust estimator. Heathcote and Silvapulle (1981) relate that the M-estimator for location from Huber (1964, p. 101) is a "close

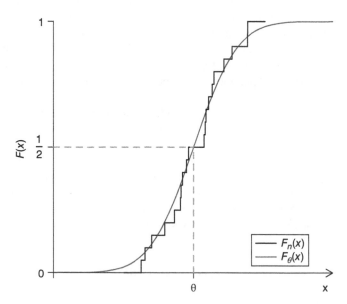

FIGURE 6.1 Example plot of the cumulative empirical distribution function generated from a cumulative normal distribution located at θ. The smooth curve is $F_{\theta}(x)$ and $F_{n}(x)$ corresponds to the step function.

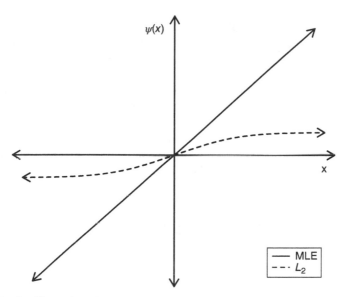

FIGURE 6.2   Illustration of boundedness of the Ψ-function for location the L_2-minimum distance estimator in comparison to the unbounded function for the maximum likelihood estimator (MLE) and at the normal location model.

competitor" to the Hodges–Lehmann estimator, which is the median of the pairwise averages $\frac{1}{2}(X_i + X_j)$ $(i < j; i, j = 1, \dots, n)$, and observe both the L_2-minimum distance estimator and the Hodges–Lehmann estimator have the same asymptotic variance at the location model. Heathcote and Silvapulle are thus spurred on to examine the estimator of both location and scale of the normal distribution, investigating its performance asymptotically even at asymmetric contaminating distributions. Hettmansperger et al. (1994) make further favorable efficiency comparisons of this L_2-minimum distance estimator for location and scale from the normal distribution at different distributions, comparing efficiency of the resultant L_2 estimator to that of the Hodges–Lehmann estimator, and the resultant scale M-estimator is compared for efficiency against the mean absolute deviation estimator, $d_n = \frac{1}{n} \sum_{i=1}^{n} |X_i - \overline{X}|$, again at different distributions in neighborhoods of the normal distribution.

The extension of Cramér–von Mises type minimum distance estimation methods to regression was almost at the same time as the appearance of the works on location and scale. Conditions for minimum Cramér–von Mises type distance estimates of regression were considered separately in Silvapulle (1981), Millar (1982), and Koul and DeWet (1983). A further summary with extensions can be found in Koul (1992). In a sense the L_2-minimum distance estimator with weight function Lebesgue measure leads in the case of the normal parametric family to the M-estimator of regression where one solves Eqs. (3.11) and (3.12) with the function $\Psi(x) = (\psi_1(x), \psi_2(x)) = (\Phi(x) - \frac{1}{2}, 1/(2\sqrt{\pi}) - \phi(x))$ [see Hettmansperger et al. (1994, p. 293)]. Hence, all the conditions for this estimator are inherited through the discussion of M-estimators. However, for more general weighting functions, such as for the MCVM that minimizes the classical Cramér–von Mises statistic where the weighting measure is the parametric distribution, this is not the case.

Beran (1977) following a lead from Matusita (1953, 1955) minimizes what is called a Hellinger distance. For continuous real-valued random variables and assuming $F_\theta(x)$ has a density $f_\theta(x)$ which is continuous, say, it is possible to estimate that density in various ways. The usual estimator is a kernel density estimator where one chooses

$$\hat{f}_n(x) = \frac{1}{n} \sum_{i=1}^{n} \hat{K}(x, X_i, h_n) \tag{6.14}$$

such that

$$\hat{K}(x, X_i, h_n) = \frac{1}{h_n} w \left(\frac{x - X_i}{h_n} \right)$$

and $w(\cdot)$ can be a symmetric nonnegative function satisfying

$$\int_{-\infty}^{\infty} w(x)dx = 1$$

and h_n is a bandwidth. A summary of kernel density estimators is proffered in Siverman (1986), for example. Here the Hellinger distance is

$$HD(\hat{f}_n, f_\tau) = 2 \int_{-\infty}^{\infty} \left(\hat{f}_n^{\frac{1}{2}}(x) - f_\tau^{\frac{1}{2}}(x) \right)^2 dx. \tag{6.15}$$

The Hellinger distance can be defined for both discrete and continuous distributions and Basu et al. (2011) make the distance a focal point for their relatively recent book. Beran (1977) describes conditions whereby if $T[G]$ is defined implicitly by

$$HD(g, f_{T[G]}) = \inf_{\tau \in \Theta} HD(g, f_\tau) \tag{6.16}$$

and for suitable densities g_n converging to g in the Hellinger distance metric (i.e. $HD(g_n, g) \to 0$ as $n \to \infty$) the functionals at those distributions satisfy for the right types of kernel density estimator, with appropriate bandwidths and appropriately defined parametric families $\mathcal{F} = \{f_\tau | \tau \in \Theta\}$, a result that $T[G_n] \to_p T[G]$ and also $\sqrt{n}(T[G_n] - T[G])$ converges in distribution to a normal random variable with mean zero and if $G = F_\theta$ the asymptotic variance is that of the inverse of Fisher information, $\mathcal{I}(\theta)^{-1}$. The fact that the estimator inherits the same asymptotic variance as that of the Cramér–Rao lower bound, achieved also by maximum likelihood estimators, with the additional advantage that the estimator is robust in Hellinger neighborhoods of the parametric distribution, set this paper as a seminal paper from which much research springs even today. It is later pointed out by Donoho and Liu (1988a) that minimum distance estimators are robust against departures in the metric distance they are defined in, whence the real question is, is the distance that is used capable of reflecting the sort of departures from the model that are realistic. Hampel (1968) captures typical departures in robustness studies according to the Prohorov metric, and for "weak" metrics such as the Cramér–von Mises distance metric, robustness over those neighborhhoods defined by that distance implies robustness over the Prohorov metric. For those integrated squared error estimators and Cramér–von Mises type estimators that are possible to be defined as M-estimators with bounded continuous Ψ functions, this robustness is inherent by the results of weak continuity of the resultant statistical functionals. On the other hand, the sorts of departures such as rounding errors and atoms of probability contamination where densities may not exist yield complications that the Hellinger neighborhoods as defined here may not cope with, at least for continuous parametric families. Donoho and Liu (1988a, 1988b) describe some further qualitative robustness results of minimum distance estimators.

The Cramér–von Mises type estimators include more general estimators such as when one minimizes with a general weight function such as when the weighting measure in Eq. (6.12), $K^*(x)$, depends also on the parameter of estimation and is other than the parametric distribution. They are no longer in general an M-estimator form when the weight function depends on the parameter, but it is insightful both historically and from the point of view of statistical inference to consider them. In summary, they have an overall form

$$\rho_{\text{CVM}}(F_n, \tau) = \int (F_n(x) - F_\tau(x))^2 \xi(x, \tau) dF_\tau(x). \tag{6.17}$$

See that the Cramér–von Mises distance estimator is defined by choosing $\xi(x, \tau) = 1$ and the L_2-minimum distance estimator of Blackman (1955) and Heathcote and Silvapulle (1981) is defined by $\xi(x, \tau) = (1/f_\tau(x))$. Interestingly, the statistic from Anderson and Darling (1952) known as the Anderson–Darling statistic which is important in the goodness of fit is given by $\xi(x, \tau) = 1/[F_\tau(x)(1 - F_\tau(x))]$. Such classical statistics are to be balanced with the general class of Kolmogorov–Smirnov statistics generated by

$$\rho_{\text{KS}}(F_n, \tau) = |F_n - F_\tau(x)| \sqrt{\{\xi(x, \tau)\}}, \tag{6.18}$$

where the usual Kolmogorov distance is given by $\xi(x, \tau) = 1$. For an empirical distribution function F_n generated by a continuous distribution G, the asymptotic critical values for a normalized statistic $\sqrt{n} d_k(F_n, G)$ can be obtained through the result

$$P(\sqrt{n} d_k(F_n, G) \le t) \to H(t) = 1 - 2\sum_{j=1}^{\infty} (-1)^{j-1} e^{-2j^2 t^2}.$$

See Kolmogorov (1933) and Smirnov (1939). Basu et al. (2011, pp. 22–24) give a succinct summary in relation to this statistic and its generalizations can be used for testing equality of two distributions for example. A natural extension is to consider the Lévy distance metric which generates the topology of weak convergence, and indeed Kozek (1998) considers a natural family of metric distances which include both the Kolmogorov distance and the Lévy distance. Here the distance is defined as

$$\rho_\alpha(F, G) = \inf\{\epsilon > 0 |\ G(x - \alpha\epsilon) - \epsilon \le F(x) \le G(x + \alpha\epsilon) + \epsilon, \quad \text{for all } x \in \mathcal{E}\}.$$

For $\alpha = 0$ we get the Kolmogorov metric (also known as the Kolmogorov–Smirnov metric), and for $\alpha = 1$ we get the Lévy metric. Kozek considers estimation of location.

Minimizing goodness of fit criteria extends more generally. In a ground breaking paper Cressie and Read (1984) introduced a unifying family of goodness of fit criteria, which were initially considered for the multinomial distribution, which incorporated several classical goodness-of-fit statistics. Basu and Lindsay (1994) refer to the continuous family analog of what they call the Cressie and Read (1984) power divergences defined as, to use the notation of Basu and Lindsay, for a model density, say $m_\beta(x)$, that is indexed by an unknown parameter β,

$$PD_\lambda(f^*, m_\beta^*) = \int f^*(x)\{[f^*(x)/m_\beta^*(x)]^\lambda - 1\}dx/\lambda(\lambda + 1)$$

$$= \int m_\beta^*(x)\{(1 + \delta^*(x))^{\lambda+1} - 1\}dx/\lambda(\lambda + 1) \qquad (6.19)$$

where they define $\delta^*(x) = (f^*(x) - m_\beta^*(x))/m_\beta^*(x)$, and $f^*(x)$ is constructed using kernel density estimation so that

$$f^*(x) = \int k(x; t, h)d\hat{F}(t),$$

where $\hat{F}$ is the empirical distribution function defined from the data and k is some smooth family of kernel functions such as the normal densities with mean t and standard deviation h. The parameter h controls the smoothness of the resulting density, with increasing h corresponding to greater smoothness [see Siverman (1986)]. For instance, Basu and Lindsay (1994) apply the same smoothing to the model so that

$$m_\beta^*(x) = \int k(x; t, h)dM_\beta(t)$$

where $M_\beta(t)$ is the model cumulative distribution function. Basu and Lindsay point out that various parameters of $\lambda = -2, -1, -1/2, 0$, and 1 generate the Neyman's

Chi-square, Kullback Leibler divergence, Hellinger distance, likelihood disparity, and Pearson's Chi-square respectively.

In a minimum divergence method, Basu et al. (1998) define a family of density-based divergence measures $d_\alpha(g,f)$ between density functions g and f to be

$$d_\alpha(g,f) = \int \left\{ f^{1+\alpha}(z) - (1 + \frac{1}{\alpha})g(z)f^\alpha(z) + \frac{1}{\alpha}g^{1+\alpha}(z) \right\} dz, \quad \alpha > 0. \quad (6.20)$$

When $\alpha = 0$ the integrand in expression (6.20) is undefined and they define

$$d_0(g,f) = \lim_{\alpha \to 0} d_\alpha(g,f) = \int g(z)\log\{\frac{g(z)}{f(z)}\}dz. \quad (6.21)$$

Here $d_0(g,f)$ is the Kullback–Leibler divergence. Basu et al. (1998) discuss the estimator that minimizes $d_\alpha(g,f_\tau)$ that in effect corresponds to a choice of an M-estimator with

$$\psi(x, \tau) = u_\tau(x)f_\tau^\alpha(x) - \int u_\tau(z)f^{1+\alpha}(z)dz,$$

where $u_t(x) = (\frac{\partial}{\partial \tau})\log f_\tau(x)$ is the maximum likelihood score function. Limiting α to zero gives the maximum likelihood estimator.

Patra et al. (2013) provide a useful connection between the power divergence families and the density power divergence families. They also note that a more general based structure for density-based divergences of the chi-squared type between the densities g and f_θ is given by

$$\rho_{CS}(G,f_\tau) = \int Q(\delta(x))f_\tau(x)dx,$$

where $\delta(x) = g(x)/f_\tau(x) - 1$ is the Pearson residual at the value x. Here Q is strictly convex and satisfies $Q(0) = 0$. Also Ghosh and Basu (2017) examine what they call an S-divergence family which depends on two parameters α and λ broadening the family of divergences. Ghosh (2014) examined the S-divergence family estimators in the case of discrete parametric models, for example.

In a paper discussing optimality properties of various distances, Markatou et al. (2016) offer some further comparisons in views of their robustness. Application of Cramér–von Mises distance estimation to actuarial data is found in Luong (2016).

For the general classes of Cramér–von Mises distances, we see that in a quest for both robustness and asymptotic efficiency, Özütrk and Hettmansperger (1997) develop a theory for a general development with

$$d_G(\tau) = \int Q(\{G(t) - F_\tau(t)\})w(t; \tau)dt,$$

where $Q \geq 0$, $Q(0) = 0$, $Q'(0) = 0$, $Q''(0) > 0$, and $w(t; \tau) \geq 0$ is a weight function. The minimizer of $d_{F_n}(\tau)$ is a generalized weighted Cramér–von Mises estimator. The authors proffer three choices for Q being

$$Q_1(t) \equiv t^2, \quad Q_2(t) \equiv \frac{t^2}{t+1}, \quad \text{and } Q_3(t) \equiv \{(t+1)^{\frac{1}{2}} - 1\}^2.$$

See that Q_1 leads to the Cramér–von Mises estimators of ρ_{CVM} above, while Q_2 and Q_3 are motivated by the Hellinger and Neyman distance functions in the context of the minimum disparity estimators of Lindsay (1994).

In a development by Scott (2001), the criterion minimized is motivated by use of a kernel density estimator $\hat{f}_n$, where one minimizes the distance

$$\int [\hat{f}_n(x) - f_\tau(x)]^2 dx,$$

which on interpretation of the integration and the use of the bandwidth yields the estimator that minimizes

$$T_{L_2E}[F_n] = \text{argmin}_\tau \int f_\tau^2(x) dx - \frac{2}{n} \sum_{i=1}^{n} f_\tau(X_i).$$

This yields an M-estimator of location and scale of the normal distribution defined by

$$\Psi(x) = [\psi_1(x), \psi_2(x)]^T = [xe^{-\frac{x^2}{2}}, (2\sqrt{2})^{-1} + (x^2 - 1)e^{-\frac{x^2}{2}}].$$

The M-estimator for location has a redescending ψ-function with an efficiency of 54.4% at the normal model. Readers should be careful to note that the reference to the integral of the squared difference between the kernel density estimator and the parametric density is sometimes also called L_2 and also *integrated squared error*, but they should not be confused with the L_2-minimum distance estimator of Blackman (1955), Heathcote and Silvapulle (1981), and Hettmansperger et al. (1994) nor the integrated squared error used in the discussion of the stable laws above, for instance, as in Heathcote (1977). Interestingly, the L_2E-estimator extends readily to the multivariate normal distribution where one minimizes

$$L_2E(\mu, \Sigma) = \frac{1}{2^p \pi^{p/2} |\Sigma|^{1/2}} - \frac{2}{n} \sum_{i=1}^{n} \phi(X_i, \mu, \Sigma),$$

where here $\mu \in \mathcal{E}^p$ and $\phi(x, \mu, \Sigma)$ is the usual multivariate normal density representing the $N(\mu, \Sigma)$ distribution.

6.2   THE L_2-MINIMUM DISTANCE ESTIMATOR FOR MIXTURES

Just as there can be more than one M-estimator for location, and in the end one has to employ an M-estimator for location for any particular application by implementing a specific ψ function, it appears that there are a myriad of estimators of parameters for general parametric models as indicated in the previous subsection and one has to choose an estimator that can be easily implemented, once programmed, on a data set. What is required is an estimator with high efficiency at the model, and a bounded continuous influence function, hopefully with good finite sample properties. By Theorem 2.6 of Huber (1981) the asymptotic breakdown point of the L_2-minimum distance estimator of location, which can loosely be described as the smallest amount in proportion of arbitrary contamination beyond which the estimating functional takes on infinite values, is shown to be the maximum possible of 0.5. These and similar types of advantages are further explored in Heathcote and Silvapulle (1981) in the estimation of location and scale of the normal distribution. It therefore is only reasonable to explore this particular estimator for the mixture model described in Clarke (1989), Clarke and Heathcote (1994), Clarke (2000b), and Clarke et al. (2017). Importantly, as is noted in those papers, the estimating functional is known to be weakly continuous and Fréchet differentiable, and hence robust. We consider here, as in the papers cited, the more general estimation of k component univariate normal mixture distributions of the form

$$F_\tau(x) = \sum_{j=1}^{k} c_j \Phi \left\{ \frac{(x-\mu_j)}{\sigma_j} \right\}. \qquad (6.22)$$

Here as throughout

$$\Phi(x) = \int_{-\infty}^{x} \phi(y)dy \ \text{ and } \ \phi(y) = \frac{1}{\sqrt{2^*\pi}} \exp\left(\frac{-y^2}{2}\right)$$

$\sum_{j=1}^{k} \epsilon_j = 1$ and $\tau \in \Theta$ is the vector of $3k-1$ parameters $\epsilon_1, \ldots, \epsilon_{k-1}, \mu_1, \ldots, \mu_k$, $\sigma_1, \ldots, \sigma_k$ which are to be estimated on the basis of the univariate sample $X_1, X_2, \ldots, X_n$. An introduction to parameterization of mixtures is afforded in the two texts by Titterington et al. (1985) and McLachlan and Peel (2000). We consider nondegenerate mixtures, where mixture parameters representing proportions of each component distribution are assumed greater than zero and component distributions are distinct with positive standard deviations. It is emphasized that we consider in this section the number of components to be known prior to estimation.

To estimate $\theta \in \Theta$ we choose to use the estimating equations that minimize the L_2 distance with weight function Lebesgue measure

$$J_n(\tau) = \int_{-\infty}^{+\infty} \{F_n(x) - F_\tau(x)\}^2 dx , \qquad (6.23)$$

which not surprisingly according to Knüsel (1969) and others lead to the M-estimating Eq. (2.1). It is perhaps instructive to determine the path for these equations in the instance of Lebesgue measure, and what one needs to do to find the exact form for the model (6.22) as succinctly explained in Clarke and Heathcote (1994). See that minimizing $J_n(\tau)$ by differentiating with respect to τ leads to a set of estimating equations which then on integrating by parts lead to

$$0 = \int_{-\infty}^{\infty} \{F_n(x) - F_\tau(x)\} \left(\frac{\partial}{\partial \tau}\right) F_\tau(x) dx$$

$$= \{F_n(x) - F_\tau(x)\} \int_{-\infty}^{x} \left(\frac{\partial}{\partial \tau}\right) F_\tau(y) dy \Big|_{-\infty}^{\infty}$$

$$- \int_{-\infty}^{\infty} \int_{-\infty}^{x} \left(\frac{\partial}{\partial \tau}\right) F_\tau(y) dy d(F_n - F\tau)(x). \qquad (6.24)$$

Checking that

$$0 = \lim_{x \to -\infty} \{F_n(x) - F_\tau(x)\} \int_{-\infty}^{x} \left(\frac{\partial}{\partial \tau}\right) F_\tau(y) dy$$

the estimating equations become

$$0 = \int_{-\infty}^{\infty} \int_{-\infty}^{x} \left(\frac{\partial}{\partial \tau}\right) F_\tau(y) dy d(F_n - F_\tau)(x)$$

$$= \frac{1}{n} \sum_{l=1}^{n} \int_{-\infty}^{X_l} \left(\frac{\partial}{\partial \tau}\right) F_\tau(y) dy - \int_{-\infty}^{\infty} \int_{-\infty}^{x} \left(\frac{\partial}{\partial \tau}\right) F_\tau(y) dy dF_\tau(x), \qquad (6.25)$$

which are of the form Eq. (2.1), with $\Psi(x, \tau)$ given by Eq. (6.13). For reasons that will become transparent, we now consider estimation of the different types of component parameters separately. Using the equations in Clarke and Heathcote (1994), this allows an expanded discussion, with some new and interesting results.

6.2.1 The L_2-Estimator for Mixing Proportions

We denote $\Psi_\epsilon(x, \tau) = (\partial/\partial\epsilon)F_\tau(x)$, where $\epsilon = [\epsilon_1, \ldots , \epsilon_{k-1}]^T$ are the free mixture component parameters, bearing in mind that $\epsilon_k = 1 - \sum_{j=1}^{k-1} \epsilon_j$. Let also for later reference $\mu = [\mu_1, \ldots , \mu_k]^T$ and $\sigma = [\sigma_1, \ldots , \sigma_k]^T$ so that $\tau^T = [\epsilon^T, \mu^T, \sigma^T]$. See then that

$$(\partial/\partial\epsilon_i)F_\tau(x) = \Phi\{\frac{(x-\mu_i)}{\sigma_i}\} - \Phi\{\frac{(x-\mu_k)}{\sigma_k}\}, \quad i = 1, \ldots, k-1$$

which then implies as according to Problem 6.3a that the quantity defined to be

$$A_{ik}(x; \boldsymbol{\mu}, \boldsymbol{\sigma}) = \int_{-\infty}^{x} (\frac{\partial}{\partial\epsilon_i})F_\theta(y)dy \qquad (6.26)$$

can be written

$$A_{ik}(x; \boldsymbol{\mu}, \boldsymbol{\sigma}) = (x - \mu_k)\left\{ \Phi\left\{ \frac{(x-\mu_i)}{\sigma_i} \right\} - \Phi\left\{ \frac{(x-\mu_k)}{\sigma_k} \right\} \right\}$$

$$+ \sigma_i\phi\left\{ \frac{(x-\mu_i)}{\sigma_i} \right\} - \sigma_k\phi\left\{ \frac{(x-\mu_k)}{\sigma_k} \right\}$$

$$+ (\mu_k - \mu_i)\Phi\left\{ \frac{(x-\mu_i)}{\sigma_i} \right\}. \qquad (6.27)$$

This is a bounded function in x, the observation space variable for compact parameter sets with component standard deviations bounded away from zero. See Problem 6.3b. Given this equation and then following Clarke and Heathcote (1994) and according to Problem 6.5 it is seen that

$$B(\boldsymbol{\mu}, \boldsymbol{\sigma}; j, i, k) = \int_{-\infty}^{\infty} \sigma_j^{-1}\phi\left\{ \frac{(x-\mu_j)}{\sigma_j} \right\} A_{ik}(x; \boldsymbol{\mu}, \boldsymbol{\sigma})dx \qquad (6.28)$$

$$= (\mu_j - \mu_k)\left\{ \Phi\left\{ \frac{(\mu_j - \mu_i)}{\sqrt{\sigma_j^2 + \sigma_i^2}} \right\} \right.$$

$$- \Phi\left\{ \frac{(\mu_j - \mu_k)}{\sqrt{\sigma_j^2 + \sigma_k^2}} \right\}\right\}$$

$$+ \sqrt{\sigma_j^2 + \sigma_i^2}\,\phi\left\{ \frac{(\mu_j - \mu_i)}{\sqrt{\sigma_j^2 + \sigma_i^2}} \right\}$$

$$- \sqrt{\sigma_j^2 + \sigma_k^2}\,\phi\left\{ \frac{(\mu_j - \mu_k)}{\sqrt{\sigma_j^2 + \sigma_k^2}} \right\} + (\mu_k - \mu_i)\Phi\left\{ \frac{(\mu_j - \mu_i)}{\sqrt{\sigma_j^2 + \sigma_i^2}} \right\}.$$

$$(6.29)$$

The following theorem defines the M-estimating equations for the mixing parameters which lead to unbiased estimates of the mixing proportions, assuming the location and scale parameters of the component populations are known and the model holds. This theorem extends the results of Clarke (1989) from the case of a single mixing proportion parameter to when there are several component proportion parameters that are to be estimated simultaneously.

Theorem 6.1: *Let ϵ be the only unknown vector parameter in the model (6.22), and let $T_l[\cdot\,; \mu, \sigma]$ be defined as a solution of the M-estimating Eq. (2.1) where the function $\Psi(x, \tau) \equiv \Psi_\epsilon(x, \tau) = [\psi_1(x, \tau), \ldots, \psi_{k-1}(x, \tau)]^T$ is defined through*

$$\psi_i(x, \tau) = A_{ik}(x; \mu, \sigma) - B(\mu, \sigma; k, i, k)$$

$$- \sum_{j=1}^{k-1} \epsilon_j \{ B(\mu, \sigma; j, i, k) - B(\mu, \sigma; k, i, k) \}, \quad i = 1, \ldots, k-1. \quad (6.30)$$

Then the M-estimator is unique, weakly continuous, and Fréchet differentiable with respect to the Kolmogorov metric and minimizes the $\mathcal{L}_2$ distance $J_n(\tau)$. In addition the M-estimator is unbiased, in that

$$E_\tau[T_l[F_n; \mu, \sigma]] = \epsilon \qquad (6.31)$$

The following is a description of the proof. It is reasonable to assume that there does not exist a nonzero vector $b_k = [b_1, b_2, \ldots, b_k]^T$ such that for all $x \in \mathcal{E}$

$$\frac{b_1}{\sigma_1} \phi \left\{ \frac{x - \mu_1}{\sigma_1} \right\} + \frac{b_2}{\sigma_2} \phi \left\{ \frac{x - \mu_2}{\sigma_2} \right\} + \cdots + \frac{b_k}{\sigma_k} \phi \left\{ \frac{x - \mu_k}{\sigma_k} \right\} = 0 \qquad (6.32)$$

since the component distributions are distinct. Hence, the $(k-1) \times (k-1)$ matrix

$$\Lambda(\mu, \sigma) = [(\lambda_{ij}(\mu, \sigma))]$$

$$= \left[\left(\int \left\{ \left(\frac{\partial}{\partial \epsilon_i} \right) F_\tau(x) \right\} \left\{ \left(\frac{\partial}{\partial \epsilon_j} \right) F_\tau(x) \right\} dx \right) \right], \quad i, j = 1, \ldots, k-1$$

is positive definite, for the reason that for arbitrary given $t = [t_1, \ldots, t_{k-1}]^T$ we have that

$$t^T \Lambda(\mu, \sigma) t = \int |t^T (\frac{\partial}{\partial \epsilon}) F_\tau(x)|^2 dx > 0 \qquad (6.33)$$

This is because for any given $t \neq 0$ there is an x_0 where $\eta(x_0; t; \mu; \sigma) \neq 0$ for

$$\eta(x; t; \mu; \sigma) = t_1 \left[\Phi \left\{ \frac{x - \mu_1}{\sigma_1} \right\} - \Phi \left\{ \frac{x - \mu_k}{\sigma_k} \right\} \right] + t_2 \left[\Phi \left\{ \frac{x - \mu_2}{\sigma_2} \right\} - \Phi \left\{ \frac{x - \mu_k}{\sigma_k} \right\} \right]$$

$$+ \ldots + t_{k-1} \left[\Phi \left\{ \frac{x - \mu_{k-1}}{\sigma_{k-1}} \right\} - \Phi \left\{ \frac{x - \mu_k}{\sigma_k} \right\} \right].$$

This would otherwise satisfy Eq. (6.32) with $b = [t^T, -(t_1 + t_2 + \ldots + t_{k-1})] \neq 0$ for all x which is a contradiction. But now see that by continuity there is a $\delta > 0$ such that $|\eta(x; t; \mu; \sigma)| > |\eta(x_0; t; \mu; \sigma)|/2$ for all $x \in [x_0 - \delta, x_0 + \delta]$, which implies Eq. (6.33) since

$$t^T \Lambda(\mu, \sigma) t > \int_{x_0 - \delta}^{x_0 + \delta} |\eta(x; t; \mu; \sigma)|^2 dx$$

$$> \int_{x_0 - \delta}^{x_0 + \delta} \left\{ \frac{|\eta(x_0; t; \mu; \sigma)|}{2} \right\}^2 dx$$

$$= \frac{\delta}{2} |\eta(x_0; t; \mu; \sigma)|^2$$

$$> 0.$$

Equation (6.30) is the result of Eq. (6.13), see also Eq. (6.25), and recognizing that

$$f_\tau(x) = \frac{\epsilon_1}{\sigma_1} \phi \left\{ \frac{x - \mu_1}{\sigma_1} \right\} + \frac{\epsilon_2}{\sigma_2} \phi \left\{ \frac{x - \mu_2}{\sigma_2} \right\} + \ldots + \frac{\epsilon_k}{\sigma_k} \phi \left\{ \frac{x - \mu_k}{\sigma_k} \right\} \quad (6.34)$$

$$= \frac{1}{\sigma_k} \phi \left\{ \frac{x - \mu_k}{\sigma_k} \right\} + \sum_{j=1}^{k-1} \epsilon_j \left\{ \frac{1}{\sigma_j} \phi \left\{ \frac{x - \mu_j}{\sigma_j} \right\} - \frac{1}{\sigma_k} \phi \left\{ \frac{x - \mu_k}{\sigma_k} \right\} \right\}, \quad (6.35)$$

whereupon taking the partial derivative with respect to ϵ the resulting equation follows from Eqs. (6.26) and (6.29).

See now that arranging the formulae for Eq. (2.1) with $\Psi_\epsilon(x, \tau)$ given by Theorem 6.1, they can be written as

$$0 = A[F_n; \mu, \sigma] + \Lambda(\mu, \sigma)\epsilon, \quad (6.36)$$

where the $(k - 1) \times 1$ vector $A[F_n, \mu, \sigma]$ has as its ith element

$$\frac{1}{n} \sum_{l=1}^{n} A_{ik}(X_l; \mu, \sigma) - B(\mu, \sigma; k, i, k).$$

Since $\Lambda(\mu, \sigma)$ is invertible given the model constraints, we have that the solution of Eq. (6.36) is in fact a sum of independent random vector variables.

$$T_1[F_n; \mu, \sigma] = -\Lambda(\mu, \sigma)^{-1} A[F_n; \mu, \sigma] = \frac{1}{n} \sum_{l=1}^{n} \iota(X_l; \mu, \sigma) \quad (6.37)$$

By the law of large numbers this converges almost surely to its expectation which equals $E_\tau[\iota(X; \mu, \sigma)]$ since they are an average of identically distributed bounded random variables. But this also equals $E_\tau[T_1[F_n; \mu, \sigma]]$. Moreover, it is easily checked (see Problem 6.4) that Ψ_ϵ satisfies Conditions **A** and subsequent Conditions **W** of Chapter 2. Therefore, it follows that Corollary 2.3 implies the estimator is robust and consistent, while Theorem 2.6 implies Fréchet differentiability. Since it is also consistent, it follows that $T_1[F_n; \mu, \sigma] \to_{a.s.} \epsilon$, whence the result of the Theorem 6.1 holds as the estimate converges to its expectation which is by consistency ϵ.

6.2.2   The L_2-Estimator for Switching Regressions

The estimating equations for the parameters in μ and σ in model (6.22) that are derived by minimizing the L_2-distance are dealt with in Clarke and Heathcote (1994). Yet there is an important body of theory that deals with the estimation of switching regressions which were defined in Quandt (1972) and made popular by Quandt and Ramsey (1978) following their introduction. They are more general than the component models that consider location only. In the latter paper, Quandt and Ramsey deal with the mixture involving the switching of two possible regressions. The idea of switching regressions when there are k regressions that could be switched is more complicated but arises when observations are given on a random variable Y and on a vector of non-stochastic regressors, when

$$Y_l = x_l^T \beta_i + u_{il} \text{ with probability } \epsilon_i$$

$$i = 1, \dots, k; \ l = 1, \dots, n$$

where $u_{il} \sim N(0, \sigma_i^2)$, and $\epsilon = (\epsilon_1, \dots, \epsilon_k)^T$ and the vectors $\beta_1, \beta_2, \dots, \beta_k$ and $\sigma_1^2, \dots, \sigma_k^2$ are unknown. We require

$$(\beta_i, \sigma_i^2) \neq (\beta_j, \sigma_j^2); \ i \neq j \text{ for all } i, j = 1, \dots, k.$$

The general distribution for the random variable Y is then

$$F_\tau(y, x) = \sum_{j=1}^{k} \epsilon_j \Phi \left\{ \frac{y - x^T \beta_j}{\sigma_j} \right\}. \tag{6.38}$$

In order to explain this model with more clarity and for purposes of illustration, in this book we consider here regressions on a single regressor variable z so that $x = (1, z)^T$. It is not necessary to do this, and indeed we could consider component regressions on several variables and also with possibly different numbers of variable regressors, but the simpler model conveys to the reader the essence of the story. Here then the parameter τ becomes

$$(\epsilon_1, \ldots, \epsilon_{k-1}, \beta_{01}, \beta_{11}, \beta_{02}, \beta_{12}, \ldots, \beta_{0k}, \beta_{1k}, \sigma_1, \ldots, \sigma_k)^T,$$

and the ith component regression has a mean of

$$\mu_i = x^T \beta_i = \beta_{0i} + \beta_{1i} z. \tag{6.39}$$

The equations for estimating ϵ are the same as in Section 6.2.1, though now note that there is a dependence on z so that x is replaced by y, z, and the component of the equations (2.1) belonging to the partial derivative with respect to ϵ of the L_2-distance becomes so that

$$\Psi_\epsilon(Y_l, z_l; \tau) = (\psi_1(Y_l, z_l, \tau), \ldots, \psi_{k-1}(Y_l, z_l, \tau))^T, \tag{6.40}$$

where

$$\psi_i(y, z, \tau) = A_{ikz}(x; \beta, \sigma) - B_z(\beta, \sigma; k, i, k)$$

$$- \sum_{j=1}^{k-1} \epsilon_j \{ B_z(\beta, \sigma; j, i, k) - B_z(\beta, \sigma; k, i, k) \}, \quad i = 1, \ldots, k-1. \tag{6.41}$$

and we make the interpretation that

$$\left. \begin{array}{l} A_{ikz}(y; \beta, \sigma) \equiv A_{i,k}(y; \mu, \sigma) \\ B_z(\beta, \sigma; j, i, k) \equiv B(\mu, \sigma; j, i, k) \end{array} \right\} \text{ but with } \mu_s = \beta_{0s} + \beta_{1s} z \\ s = 1, \ldots, k$$

$$i = 1, \ldots, k-1.$$

Now the argument for finding the minimizing equations that correspond to partial derivatives with respect to β of the L_2 distance are entwined with the arguments of Clarke and Heathcote (1994) for estimating μ in the model (6.22). Problem 6.2 which requires one to prove identities (2.3) and (2.4) of that paper can be used to show, for example,

$$E_\tau \left[\int_{-\infty}^X (\frac{\partial}{\partial \mu_i}) F_\tau(y) dy \right] = \epsilon_i C_{ik}(\tau); i = 1, \ldots, k-1 \tag{6.42}$$

where

$$
C_{ik}(\tau) = \begin{cases}
-\Phi\left\{\dfrac{(\mu_k-\mu_i)}{\sqrt{\sigma_k^2+\sigma_i^2}}\right\} \\
+\sum_{j=1}^{k-1}\epsilon_j[\Phi\left\{\dfrac{(\mu_k-\mu_i)}{\sqrt{\sigma_k^2+\sigma_i^2}}\right\} \\
-\Phi\left\{\dfrac{(\mu_j-\mu_i)}{\sqrt{\sigma_j^2+\sigma_i^2}}\right\}]
\end{cases}
\tag{6.43}
$$

and

$$
E_\tau\left[\int_{-\infty}^{X}(\frac{\partial}{\partial\sigma_i})F_\tau(y)dy\right] = \epsilon_i D_{ik}(\tau); i = 1, \ldots, k-1
\tag{6.44}
$$

where

$$
D_{ik}(\tau) = \begin{cases}
\left\{\dfrac{\sigma_i}{\sqrt{\sigma_i^2+\sigma_k^2}}\right\}\phi\left\{\dfrac{(\mu_i-\mu_k)}{\sqrt{\sigma_i^2+\sigma_k^2}}\right\} \\
+\sum_{j=1}^{k}\epsilon_j\left[\left\{\dfrac{\sigma_i}{\sqrt{\sigma_i^2+\sigma_j^2}}\right\}\phi\left\{\dfrac{(\mu_i-\mu_j)}{\sqrt{\sigma_i^2+\sigma_j^2}}\right\}\right. \\
\left.-\left\{\dfrac{\sigma_i}{\sqrt{\sigma_i^2+\sigma_k^2}}\right\}\phi\left\{\dfrac{(\mu_k-\mu_i)}{\sqrt{\sigma_i^2+\sigma_k^2}}\right\}\right]
\end{cases}.
\tag{6.45}
$$

Fitting the model (6.22) as in Clarke and Heathcote (1994), the estimating equations for the component distribution location parameters are governed by the ψ-functions

$$
\psi_{i'}(x,\tau) = -\Phi\left\{\frac{x-\mu_i}{\sigma_i}\right\} - C_{ik}(\tau);
$$

$$
i' = k, \ldots, 2k-1; i = i' - k + 1.
$$

It is then not hard to see that for the linear switching regressions model (6.38) using Eq. (6.39) the estimating equations for β are defined by ψ-functions

$$
\begin{pmatrix}\psi_{i',1}(y,z,\tau) \\ \psi_{i',2}(y,z,\tau)\end{pmatrix} = \left[-\Phi\left\{\frac{y-\mu_i}{\sigma_i}\right\} - C_{ikz}(\tau)\right]\begin{pmatrix}1 \\ z\end{pmatrix};
$$

$$
i' = k, \ldots, 2k-1;
$$

$$
i = i' - k + 1
$$

$$
\text{where } C_{ikz}(\tau) \equiv C_{ik}(\tau)
$$

$$
\text{with } \mu_i = \beta_{0i} + \beta_{1i}z
\tag{6.46}
$$

and the estimating equations for the σ are governed by ψ-functions arguing the same as in Clarke and Heathcote (1994) so that

$$\psi_{i'}(y, z, \tau) = \phi \left\{ \frac{y - \mu_i}{\sigma_i} \right\} - D_{ikz}(\tau);$$

$$i' = 2k, \ldots, 3k - 1; i = i' - 2k + 1$$

$$D_{ikz}(\tau) \equiv D_{ik}(\tau) \text{ with } \mu_i = \beta_{0i} + \beta_{1i}z. \tag{6.47}$$

The M-estimating equations for the model (6.38) using (6.39) are then

$$\frac{1}{n} \sum_{l=1}^{n} \Psi(Y_l, z_l, \tau) = \frac{1}{n} \sum_{l=1}^{n} \begin{pmatrix} \Psi_\epsilon(Y_l, z_l, \tau) \\ \Psi_\beta(Y_l, z_l, \tau) \\ \Psi_\sigma(Y_l, z_l, \tau) \end{pmatrix} = \mathbf{0}, \tag{6.48}$$

and the functions are defined with Ψ_ϵ as in Eq. (6.40) and

$$\Psi_\beta = \left(\psi_{k,1}, \psi_{k,2}, \psi_{(k+1),1}, \psi_{(k+1),2}, \cdots, \psi_{(2k-1),1}, \psi_{(2k-1),2} \right)^T$$

$$\Psi_\sigma = \left(\psi_{2k}, \ldots, \psi_{3k-1} \right)^T.$$

Here then the vector function Ψ is of dimension $4k - 1$.

6.2.3   An Example Application of Switching Regressions

The following experiments were carried out by a chemistry honors student at Murdoch University in Semester 2 2015. Each experiment involved determining the diameter of carbon-based microspheres using scanning electron microscopy. He had synthesized those spheres using a hydrothermal technique where solutions of glucose, water, and different organic additives are reacted at $180°C$ in a small autoclave (i.e. under pressure). The aim was the formation of a polymeric carbon-based material that tends to form discrete spheres with diameters in the low micron to nanometer range and to see if there were any significant changes based on the type of additive used. An experiment involved the measures of diameters in a kiln taken at various concentrations. Each experiment was carried out independently resulting in one recording of diameter in the variable "Diam" and one recording of the concentration variable "Conc" each time. There were four different concentrations used in all of the experiments with different numbers of observations at each different concentration. Results of the experiments are recorded in Table 6.1.

Side-by-side histograms of the data are exhibited in Figure 6.3. Curiously, none of the histograms appeared normal, in fact they all appeared bimodal. The student sought answers to questions about the possibility of there being two regression lines, whether the regression line of diameter versus concentration ("Conc") passing through the component population with the largest mode could be significant,

TABLE 6.1   Diameters at Different Concentrations

Conc = 0.26 mol/l		Conc = 0.51 mol/l		Conc = 0.77 mol/l		Conc = 1.03 mol/l	
2.2	7.6	2.8	7.0	2.4	8.4	2.0	7.8
2.3	7.9	2.8	7.1	2.7	8.5	2.5	7.9
2.4	8.0	3.0	7.2	2.8	8.5	3.2	8.1
2.6	8.1	3.2	7.3	5.8	8.5	3.4	8.1
2.8	8.1	3.3	7.3	6.6	8.5	3.4	8.2
3.5	8.1	3.5	7.4	6.7	8.9	3.7	8.3
5.7	8.1	3.6	7.6	6.8	9.0	3.8	8.4
6.5	8.1	3.9	7.6	7.4	9.0	5.7	8.6
6.6	8.2	4.9	7.6	7.4	9.1	5.7	8.7
6.6	8.3	5.6	7.8	7.6	9.1	6.4	8.7
7.0	8.6	6.0	7.9	7.7	9.7	6.5	8.9
7.1	8.6	6.1	7.9	7.8	10.1	6.5	8.9
7.1	9.2	6.2	8.0	7.8		6.5	9.2
7.3	9.2	6.5	8.5	8.0		6.6	9.2
7.4		6.5	8.7	8.1		7.2	9.3
7.4		6.6	9.0	8.1		7.2	9.4
7.4		6.6	9.3	8.2		7.5	9.6

and whether this was true also of the potential regression line passing through the unexpected components corresponding to the smaller mode? Clearly regarding the data as a single simple linear regression with normal errors here is dubious but one **can** on the other hand employ switching regressions. The variable z in the model (6.39) is concentration. As expected at each level of concentration, z, there are two possible component populations with means $\beta_{01} + \beta_{11}z$ and $\beta_{02} + \beta_{12}z$ respectively. The aim is to estimate the switching regression parameters and test whether β_{1i} is significantly different from zero. This can be attained, for example, by using an asymptotic confidence interval based around the regression estimate and a standard error drawn from the jackknifed estimate of covariance of the parameters. From several different starting points the algorithm converged to the same point of the parameter space at $(\hat{e}_1, \hat{\beta}_{01}, \hat{\beta}_{02}, \hat{\beta}_{11}, \hat{\beta}_{12}, \hat{\sigma}_1, \hat{\sigma}_2) = (0.7769, 7.3637, 3.3438,$ $0.7685, 0.0863, 0.9737, 0.9762)$. From the estimates and their jackknifed variance estimates one can get estimates of standard errors for the estimated slopes

$$\hat{\beta}_{11} \pm s.e.(\hat{\beta}_{11}) = 0.7685 \pm \sqrt{0.063\,43} = 0.7685 \pm 0.2518$$

$$\hat{\beta}_{12} \pm s.e.(\hat{\beta}_{22}) = 0.0863 \pm \sqrt{0.103\,05} = 0.0863 \pm 0.3210$$

The conclusion is that the fitted line through the major proportion of data is marginally significant, though the line fitted through the smaller proportion of data is perhaps not significant in any way. The plot of data with fitted switching regressions is given in Figure 6.4.

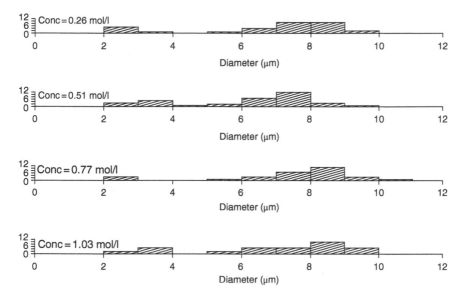

FIGURE 6.3   Side-by-side plots of histograms at each of the four concentration levels, 0.26, 0.51, 0.77, and 1.03 mol/l respectively.

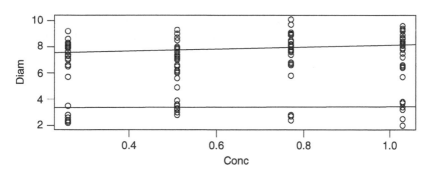

FIGURE 6.4   Plot of switching regressions.

6.3   OTHER MINIMUM DISTANCE ESTIMATION APPLICATIONS

The model F_θ involved in M-estimation and/or minimum distance estimation is not just limited to normal distributions and mixtures of normal distributions. The application is indeed wide and encompasses a multitude of estimation problems. As examples we provide a discussion of two types of quite complex models that the author has been involved with in theory and practice.

6.3.1 Mixtures of Exponential Distributions

One such general class of complex models investigated in collaboration by the author include the study of mixtures of exponential distributions. There are real biological structures and mechanisms that can be interpreted by such models and that can give understanding in the way, for example, that the inner working of biological cells behave. Electrical signals traveling through biological cell membranes have principally been studied through the passage of ions traveling through a gating mechanism known in the literature as an ion channel. See Sigg (2014) for a general introduction to the ion channel biology. Statistical modeling and inference for a single-ion channel has principally been carried out using finite-state space continuous-time Markov chains. See Colquhoun and Hawkes (1982), Ball et al. (1999), and Sakmann and Neher (1995) for an overview. See also Ball and Rice (1992) for a bibliography of Markovian models in ion channel kinetics. An electrical signal marked by the passage of ions through the cell membrane indicates that the ion channel is open. When the signal in terms of electric current stops the ion channel is essentially closed and so no ions are passing through. The duration with which the ion channel remains closed(open) can be modeled by assuming there are one or more closed(open) states linked in terms of a finite state continuous time Markov chain. The duration with which the ion channel remains closed(open) relies on the time spent in successive closed(open) states of the Markov chain. It is a feature of this modeling that the resultant closed(open) dwell times have as their distribution mixtures of exponential distributions. Times spent in any individual state are naturally exponential, whereas the accumulated time in a set of closed states can be a mixture, as explained by an analysis of the finite state continuous time Markov chain model. For example, consider the structures of Figures 6.5 and 6.6 that may represent possible systems, with different numbers of closed states.

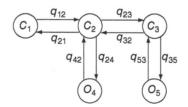

FIGURE 6.5   Five-state model.

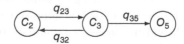

FIGURE 6.6   Three-state model.

Here q_{ij} are the transition rates from state i to state j, where $d_i = \sum_{j \neq i} q_{ij}$ is the transition rate out of state i. Let $p_{ij} = \frac{q_{ij}}{d_i}$ which implies $\sum_{j \neq i} p_{ij} = 1$. A single sojourn in state i has an exponential distribution with Laplace transform $\varphi_i(s) = \frac{d_i}{d_i + s}$ ($s \geq 0$). This is the Laplace transform of an exponential random variable with mean $1/d_i$.

As explained in Clarke and McKinnon (2005), models such as the five-state model can lead to an equilibrium distribution for closed dwell times of a three component exponential mixture distribution, and a more simple three-state model gives an equilibrium distribution of closed times as a mixture of two exponentials. More generally, it is possible subject to some conditions, to contemplate a model for the distribution of the length of stay in the closed states that is a mixture of k-exponential distributions.

$$F_\theta(x) = \sum_{j=1}^{k} \epsilon_j G(x; \lambda_j) = \sum_{j=1}^{k} \epsilon_j (1 - e^{-\lambda_j x})$$

$$= 1 - \epsilon_1 e^{-\lambda_1 x} - \cdots - \epsilon_{k-1} e^{-\lambda_{k-1} x} - (1 - \epsilon_1 - \cdots - \epsilon_{k-1}) e^{-\lambda_k x} \quad (6.49)$$

where we have the constraints $\sum_{j=1}^{k} \epsilon_j = 1$, and (ϵ_j, λ_j) are both positive, $j = 1, \dots, k$; $\lambda_i \neq \lambda_j$ for $i \neq j$. The vector parameter $\theta = (\epsilon_1, \dots, \epsilon_{k-1}, \lambda_1, \dots, \lambda_k)^T$ here. Correspondingly, the densities for the five-state model and the three-state model are respectively

$$f_\theta(t) = \epsilon_1 \lambda_1 e^{-\lambda_1 t} + \epsilon_2 \lambda_2 e^{-\lambda_2 t} + \epsilon_3 \lambda_3 e^{-\lambda_3 t}, \quad (6.50)$$

$$f_\theta(t) = \epsilon_1 \lambda_1 e^{-\lambda_1 t} + \epsilon_2 \lambda_2 e^{-\lambda_2 t}. \quad (6.51)$$

The equations for minimizing the L_2 distance are again of the M-estimating form and the form of the defining psi function is given in Appendix A of Clarke and McKinnon (2005). It is a bounded smooth psi function and again conditions for Fréchet differentiability and weak continuity of the estimating functional are met, indicating robustness of the L_2 minimum distance estimator. This is not the case for the corresponding maximum likelihood estimator since the efficient score function is in that case again unbounded. This can be easily checked, for example, in the case of only one component distribution. It is also true for general k, because of the exponentially decreasing tail of the mixture density. For the particular models (6.50) and (6.51), there is a nonlinear relationship between the parameter θ and the transition rates of the respective models. This is exploited by Clarke and McKinnon by first examining the performance of the L_2 estimates of θ in comparison to the maximum likelihood estimator and then translating that to the performance in terms of the estimation of the transition rates for the respective models.

The three-state model is a special case of the five-state model. For instance, the model in Figure 6.6 is a special case of the five-state model of Figure 6.5, putting transition rates $q_{21} = q_{24} = 0$. Here q_{ij} are transition rates (per second) from state i to state j. States $C_1 - C_3$ form the closed class of states and O_4–O_5 represent the open states. It is a feature of many robustness studies to consider, for example, the behavior of an estimation procedure in some small departure from the assumed model. Clarke and McKinnon (2005) consider parametric estimation of the closed dwell time distributions that would arise in the model as described by Figure 6.6. To see how robust is the estimation procedure, they consider what happens if in fact the underlying model is given in Figure 6.5 in which case there are two extra states C_1 and O_4. They compare the performance of the estimators when in fact there are two extra states: one closed and one open, for which there is an underlying three-component mixture for the equilibrium marginal distribution of closed-time sojourns, though the third component distribution has a small proportion but a large mean. In behaving as being unaware of the extra states, the L_2 minimum distance estimator found by fitting a three-component mixture model to independent realizations from Eq. (6.50) outperforms the solution obtained from the EM algorithm. This also has relevance to the actual transition rates derived from the estimated mixtures. For example, the relevant change in the underlying distribution considers modeling the closed-time distribution in the five-state model. These distributions mimic only slight departures from the original distributions of the entertained three-state model in Figure 6.6, for instance when q_{21}, $q_{24} \ll q_{23}$, q_{32}. For example, a transition to the state C_1 when the ion channel is spending a sojourn in the class of closed states occurs infrequently; however, when it occurs (at least for the transition rates entertained in that paper) the typical stay is on average longer than for a visit to any other of the other two closed states. The effect noted there is to yield an underlying closed dwell time distribution which is a three-component mixture of exponentials but with one component having a small proportion with a large mean.

In correspondence with Professor Frank Ball, he pointed out that the actual closed times for the five-state model given here are dependent. Standard errors of estimates would in fact, therefore, have to be larger than those reported in Clarke and McKinnon (2005). A model in which closed dwell times would be independent and for which the closed dwell times follow a three-component model would be, for example, $C_1 \Rightarrow C_2 \Rightarrow C_3 \Rightarrow O_5$.

Now that we have described the relevance of mixtures of exponential distributions to the expansive field of finite-state space continuous-time Markov chains as applied to ion channel models, we shall desist further from relating to the specific applications. Relationships with the finite-state space continuous-time Markov chains have been related in Clarke and McKinnon (2005). Rather than involving the choices of transition rates, we simply illustrate here the features to do with modeling the two distributions, the three-component distribution of Eq. (6.50) and the two-component distribution of Eq. (6.51), in some further simulations to illustrate to the reader the main point. The applications of finite mixtures of exponential distributions extend to other fields where mixtures of exponential distributions are

TABLE 6.2   Parameters and Means of Sample Estimates

	Parameter Set (1)			Parameter Set (2)			Parameter Set (3)		
	ϵ_1	λ_1	λ_2	ϵ_1	λ_1	λ_2	ϵ_1	λ_1	λ_2
True	0.6	1.0	20	0.6	1.5	150	0.15	250	5000
L_2	0.599	0.999	20.19	0.597	1.491	150.7	0.153	255.9	5047.0
EM	0.596	1.015	20.11	0.5661	65.45	133.4	0.152	251.4	5041.0

TABLE 6.3   Root Mean Squared Error

	Parameter Set (1)			Parameter Set (2)			Parameter Set (3)		
	ϵ_1	λ_1	λ_2	ϵ_1	λ_1	λ_2	ϵ_1	λ_1	λ_2
L_2	0.024	0.044	2.045	0.021	0.066	13.82	0.014	28.59	218.90
EM	0.037	0.100	3.215	0.099	245.00	48.46	0.012	8.56	215.60

important which are many. For more applications, see Jewell (1982) for a classic work and Cai (2004) for a relatively recent work in connection with insurance. Firstly, Table 6.2 illustrates that the L_2-minimum distance estimator can do well and even better than the EM algorithm when applied to a mixture of two exponential distributions. Using $m = 100$ simulations of samples of size $n = 1\,000$ the root mean squared errors gained for the three parameter sets of $\theta = (\epsilon_1, \lambda_1, \lambda_2)$ shows that even in large samples the L_2-estimator can compete well with the maximum likelihood estimator as implemented via the EM algorithm. Indeed for the parameter set (2) in Table 6.2 there were several cases of maximum likelihood estimates that, while converging, were still far from the bulk of the remaining values, giving as a result a larger than expected root mean squared error, particularly so for the parameter λ_1 (Table 6.3). Nevertheless, the histograms of component parameter L_2 estimates were exhibiting normality associated with the asymptotic behavior. As a measure of consistency, the averages of the L_2 estimates were close to the true parameters. However, in the case of fitting a mixture of two exponential distributions when there may be a small amount say 1% of contamination from a third exponential distribution with a much larger mean, the L_2 estimator was at least stable in the presence of such contamination. On the other hand, the EM algorithm failed to converge on any occasion with the contaminating distribution. The root mean squared errors for the L_2 estimator are given in Table 6.4.

6.3.2   Gamma Distributions and Quality Assurance

This section is with permission based on the article of Clarke et al. (2012). The general form of the gamma density takes α as the shape parameter and β as the scale parameter. Thus, the distribution is defined by the density

$$f_{\alpha,\beta}(x) = \frac{1}{\Gamma(\alpha)\beta^\alpha} x^{\alpha-1} \exp\left\{ \frac{-x}{\beta} \right\} \quad ; x > 0; \; \alpha, \beta > 0, \tag{6.52}$$

TABLE 6.4   Root Mean Squared Error for a 1% Contaminated Model:
$F = 0.99F_\theta(x) + 0.01F_{\lambda_3}(x)$. **Here** $\theta = (\epsilon_1, \lambda_1, \lambda_2)$

| | $\lambda_3 = 0.01$ | | | $\lambda_3 = 0.001$ | | | $\lambda_3 = 1$ | |
| | Parameter Set (1) | | | Parameter Set (2) | | | Parameter Set (3) | |
	ϵ_1	λ_1	λ_2	ϵ_1	λ_1	λ_2	ϵ_1	λ_1	λ_2
L_2	0.031	0.09724	2.702	0.0241	0.132	18.67	0.0214	84.480	364.300

where $\Gamma(\alpha) = \int_0^\infty y^{\alpha-1} \exp\{-y\}dy$ is the gamma function. The expected value of the distribution is given by $\mu = \alpha\beta$ and the variance by $\sigma^2 = \alpha\beta^2$. Here the parameter $\theta = (\alpha, \beta)^T$ and $F_\theta(x)$ is the cumulative parametric distribution function. Special cases of the gamma distribution include the exponential distribution given by the density with $\theta = (1, 1/\lambda)^T$ and the chi-squared distribution with r degrees of freedom which is given by $\theta = (r/2, 2)$. We abbreviate this distribution to χ_r^2. The classical methods of estimation include the method of moments and the maximum likelihood estimator. The maximum likelihood estimator for the gamma parametric model is investigated in detail in Bowman and Shenton (1988), yet many have preferred the simpler method of moments estimator. Using the first two moments, one obtains the straightforward estimates $(\hat\alpha, \hat\beta) = (\bar{x}^2/s^2, s^2/\bar{x})$, where $\bar{x}$ is the sample mean and s^2 is the usual unbiased sample variance estimate of σ^2, the population variance. The estimate for the mean is then simply $\hat\mu = \bar{x}$ which is then the usual sample mean. Both the method of moments and the maximum likelihood estimator have unbounded influence functions at the gamma distribution and are not robust. A detailed comparison of these and the B-optimal M-estimators of Hampel et al. (1986) implemented in Marazzi and Ruffieux (1996), the L_2 estimator and the Cramér–von Mises distance estimator minimizing $\omega^2(\tau)$, as in Eqs. (6.23) and (6.5), is given in Clarke et al. (2012). The latter estimators are all noted to be weakly continuous and Fréchet differentiable. The bounded nature of their respective influence functions at the gamma distribution are reflected in the sensitivity curves illustrated in Clarke et al. (2012) and given here in Figure 6.7. Briefly, referring to Hampel et al. (1986, p. 93), Tukey's *sensitivity curve* (SC) is defined by

$$SC_n(x) = \frac{\{T[(1 - \frac{1}{n})F_{n-1} + \frac{1}{n}\delta_x] - T[F_{n-1}]\}}{1/n}$$

Here F_{n-1} is the empirical distribution function of $(x_1, x_2, \ldots, x_{n-1})$, δ_x is the cumulative distribution with all its mass at the point x, and T is a functional such that $\hat\theta = T[F_n]$. As in Clarke et al. (2012) rather than basing the curves on actual samples, $(x_1, \ldots, x_{n-1})$ the "stylized" sensitivity curves have been obtained from artificial samples with observations $x_i = F_\theta^{-1}(i/n)$. The example data of Clarke et al. (2012) is taken from a real data problem motivated by the blending of ores

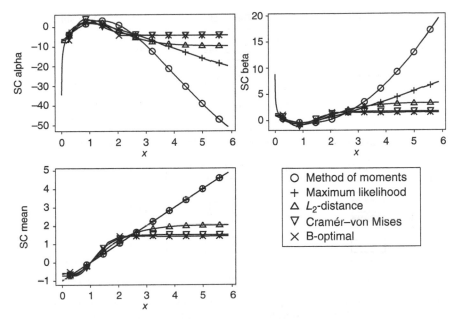

FIGURE 6.7   Sensitivity curves based on stylized samples ($n = 50$) with $\alpha = 2$ and $\beta = 0.5$. (*Source*: Clarke et al. (2012). Reproduced with the permission of Springer Nature.)

at a bauxite refinery in Western Australia. In summary, the motivation behind the analysis was to automate the statistical analysis of data for which there was reason to believe the data would follow a gamma distribution, but because of the occasional aberration in the X-ray equipment for examining powdered rock in a production line setting, there subsequently appeared outliers in some samples taken. Reasonably large samples could be expected to be analyzed in the online measurement process (Eyer and Riley, 1999) and the objective was to gain knowledge of mean concentration levels of ore on a continuing basis. It was recommended that because of both robustness and the relative ease of which the Cramér–von Mises distance could be calculated, thus leading to computationally efficient estimation of parameters of fitted distributions, that the Cramér–von Mises estimator would be a worthy candidate for use. Asymptotic relative efficiencies were also compared in Clarke et al. with a favorable comparison between the Cramér–von Mises estimator (MCVM) and the B-optimal M-estimators. It was also noted in that paper that the optimal tuning parameters of the B-optimal estimator were parameter dependent. Figure 6.8 refers to the box and whisker plots of two summary measures called representation indicators, the Mahalanobis distance and residual ratios, taken from three different data sets being of sample size $n = 297$, $n = 297$, and $n = 74$ respectively. Data set 3 has two extreme

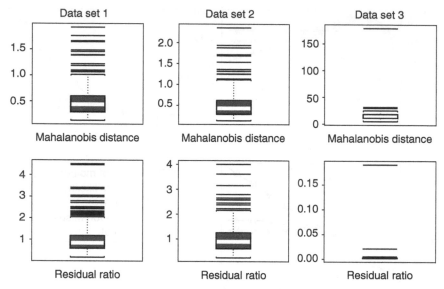

FIGURE 6.8 Box and whisker plots of Alcoa representation indicator data. (*Source*: Clarke et al. (2012). Reproduced with the permission of Springer Nature.)

outliers [Agostinelli et al. (2014, Example, p. 98)]. The fitted density curves to histograms of the data are all in relative agreement for Data sets 1 and 2 for all five estimators, whereas the method of moments estimator and the maximum likelihood estimator lead to fitted densities that are in broad disagreement with the histograms for the respective measures in Data set 3. See Figure 6.9. On the other hand, the weakly continuous Fréchet differentiable estimators of the MCVM, B-optimal, and L_2 estimators give relatively good fits to the histograms for the bulk of the data for the respective summary measures in Data set 3.

Further work with the Cramér–von Mises estimator with interesting applications has been carried out for the related beta distribution. See Thieler et al. (2016). In addition, computer algorithms are available for the Cramér–von Mises estimates for several distributions. See Kohl and Ruckdeschel (2010) and the references cited therein.

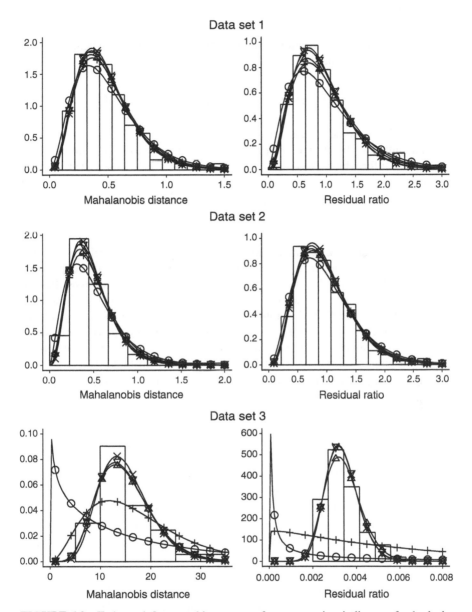

FIGURE 6.9   Estimated fits over histograms of representation indicators for both the Mahalanobis distances and the residual ratios of data sets 1–3. (*Source*: Clarke et al. (2012). Reproduced with the permission of Springer Nature.)

PROBLEMS

6.1. The R-estimator of location is defined by the functional T, where $T[F]$ is a solution of the equation

$$\int_0^1 J\left\{\frac{1}{2}[s+1-F(2T[F]-F^{-1}(s))]\right\} ds = 0.$$

Here $J(s)$ is some weight function satisfying $J(1-s) = -J(s)$, $0 < s < 1$. The influence function of T at F, where F is defined to be a symmetric distribution for which $T[F] = 0$, and F is such that it has density $f(x)$, is given by

$$\text{IF}(x, F, T) = \frac{J(F(x))}{\int_{-\infty}^{+\infty} J'(F(x))f(x)^2 dx}.$$

(a) Given an *i.i.d.* random sample $X_1, \ldots, X_n$ from a distribution F, the Hodges–Lehmann estimator of location is given by the expression $\text{med}\{(X_i + X_j)/2\}$, that is, the median of all pairwise averages. In fact, this estimator can also be written in the form of an R-estimator as $T[F_n]$ where F_n is the empirical distribution function and T is governed by the choice of $J(t) = t - 1/2$. Calculate the asymptotic variance of this estimator at the standard normal distribution. Hence, what is its asymptotic efficiency (at the standard normal distribution)?

(b) Consider the minimum distance estimator of location defined at that parameter $\hat{\mu}_n = \tilde{T}[F_n]$ that minimizes

$$J_n(\mu) = \int_{-\infty}^{+\infty} [F_n(x) - F(x - \mu)]^2 dx.$$

It is a solution of $\frac{\partial}{\partial \mu} J_n(\mu) = 0$. Assuming interchange of partial differentiation and integration and then using integration by parts, show that $\hat{\mu}_n$ is an M-estimator of location defined by the function $\psi(x) = F(x) - \frac{1}{2}$.

(c) Give the influence function of the M-estimator defined in part (b) of this question, again assuming $\tilde{T}[F] = 0$. How does it compare with the influence function for the Hodges–Lehmann estimator?

(d) From Huber (1981, Example 4.4, p. 67) the asymptotic breakdown point for the Hodges–Lehmann estimator is approximately 0.293. From Huber (1981, Theorem 2.6, p. 54) it can be deduced the equivalent M-estimator has asymptotic breakdown point of one half. By comparing the properties of the minimum distance estimator and the Hodges–Lehmann estimator, which of the two estimators would you prefer if you were informed that the data you wish to analyze are from a possibly contaminated normal distribution? Give reasons.

6.2. Establish the following identities which are Eqs. (2.3) and (2.4) of Clarke and Heathcote (1994):

(a) $\int_{-\infty}^{\infty} \phi\left\{(x - \mu_i)/\sigma_i\right\} \phi\left\{(x - \mu_j)/\sigma_j\right\} dx =$
$$\left\{(\sigma_i\sigma_j)/\sqrt{\sigma_i^2 + \sigma_j^2}\right\}\phi\{(\mu_i - \mu_j)/\sqrt{\sigma_i^2 + \sigma_j^2}\}.$$

(b) $\int_{-\infty}^{\infty} \frac{1}{\sigma_j}\phi\left\{(x - \mu_j)/\sigma_j\right\} \Phi\left\{(x - \mu_i)/\sigma_i\right\} dx = \Phi\{(\mu_j - \mu_i)/\sqrt{\sigma_i^2 + \sigma_j^2}\}.$

6.3. (a) Establish Eq. (6.27) by using integration by parts and the Eq. (6.26) for the univariate mixture of normal distributions model.

(b) Show by taking limits of the observation space variable to minus and plus infinity that, for $\theta \in D$ where D is a compact subset of the parameter space, the function $A_{ik}(x; \boldsymbol{\mu}, \boldsymbol{\sigma})$ is a bounded function in the observation space variable.

6.4. Check assumptions **A** are satisfied for the function $\boldsymbol{\Psi}_\epsilon$ with respect to neighborhoods generated by the Kolmogorov distance metric and using the normal mixture parametric model (6.22). Check also conditions **W** are also satisfied.

6.5. Establish using Eq. (6.28) the result (6.29) using the identities (6.2a) and (6.2b) of the problem above.

6.6. Show that the elements of the matrix $\Lambda(\mu, \sigma)$ are given by $\lambda_{ij}(\mu, \sigma) = -(B(\boldsymbol{\mu}, \boldsymbol{\sigma}; j, i, k) - B(\boldsymbol{\mu}, \boldsymbol{\sigma}; k, i, k))$.

7

L-ESTIMATES AND TRIMMED LIKELIHOOD ESTIMATES

7.1   A PREVIEW OF ESTIMATION USING ORDER STATISTICS

In this section we deal briefly with statistics based on linear combinations of order statistics. L-statistics as they are called have been dealt with in detail in earlier books on robustness, even as early as Huber (1977, 1981), and they were well known by the appearance of Hampel et al. (1986), yet they motivate much research even today and a recent summary can be found in Jurečková et al. (2012). The best-known L-statistic is the trimmed mean and understanding this statistic in the univariate estimation of location has motivated legions of robustness followers. The 5% trimmed mean removes approximately the 5% largest and the 5% smallest observations in a sample before taking the average of the remainder. Thus, in a sample of 20 observations the largest and the smallest would be removed before taking the mean of the remaining 18 observations. We focus on the trimmed mean rather than other linear combinations of order statistics for its natural extensions to the combination of L- and M-estimation that follow in subsequent derivations with the advent of trimmed likelihood methodology. By realizing the limitations of what is known as the adaptive trimmed mean, where one chooses to trim a proportion of data so that the variance of the trimmed mean subsequently is minimized, the relative success of the adaptive trimmed likelihood mean discussed

Robustness Theory and Application, First Edition. Brenton R. Clarke.
© 2018 John Wiley & Sons, Inc. Published 2018 by John Wiley & Sons, Inc.
Companion website: www.wiley.com/go/clarke/robustnesstheoryandapplication

in later chapters leads to a natural approach where one identifies outliers in data. The trimmed mean is a weakly continuous Fréchet differentiable statistic at the normal distribution (see Huber, 1996, pp. 18–19 and Boos 1979) and a naturally robust statistic. In a sense then, since the trimmed mean has the same general form of influence function as Huber's minimax M-estimator of location Huber (1996, Example 3, p. 17) and Maronna et al. (2006, section 10.7), it is an appropriate statistic to use if we believe our potential contamination will be symmetric and not be in a proportion in excess of the overall trimming amount. But contamination is usually never symmetric; certainly, this is the case in finite samples, and many authors following the appearance of Rousseeuw and Leroy (1987) argue that the proportion of contamination can be as much as up to approximately one half of the data. This in effect could be interpreted if we were to use a trimmed mean that one should trim approximately one half the data from each tail, ultimately leaving us with the median, a not very efficient statistic at the normal distribution, and a statistic that is not Fréchet differentiable, it being a quantile. The other limiting case is to trim 0% per cent from each tail which would leave us with the sample average, a clearly non-robust statistic.

Finding an alternative to the use of the sample mean and the associated $100\gamma\%$ Student's t-confidence interval given by

$$\bar{x} \pm t_{n-1}\left(\frac{1-\gamma}{2}\right)\frac{s_n}{\sqrt{n}} \tag{7.1}$$

was a motivation for the seminal paper by Tukey and McLaughlin (1963). While Student's t has a moderate "robustness of validity," its power, "robustness of efficiency," and hence also the length of confidence intervals and size of standard errors are very non-robust. Cressie (1980) provides a survey on the performance of the Student's t-statistic. Staudte and Sheather (1990, p. 156) advocated the use of the alternative tailored t which is formed from the trimmed mean

$$y_{T_g} = \frac{1}{n-2g}\left(y_{g+1} + y_{g+2} + \cdots + y_{n-g}\right) = \frac{\sum_{T_g} y_j}{\sum_{T_g} 1} \tag{7.2}$$

where the n ordered observations are

$$y_1 \le y_2 \le \cdots \le y_n. \tag{7.3}$$

A square root of a suitably scaled Winsorized sum of squares of deviations was used for a denominator. A Winsorized mean, for example, is given by

$$y_{W_g} = \frac{1}{n}\left(g y_{g+1} + y_{g+1} + y_{g+2} + \cdots + y_{n-g} + g y_{n-g}\right) = \frac{\sum_{W_g} y_j}{n}, \tag{7.4}$$

which is also an L-statistic. Staudte and Sheather (1990) and Tukey and McLaughlin (1963) both quote numerical studies in support of the approximate validity of nominal $100\gamma\%$ confidence intervals for the mean; these intervals are of the form

$$y_{T_g} \pm t_{h-1}\left(\frac{1-\gamma}{2}\right)\sqrt{\frac{\text{SSD}_{W_g}}{h(h-1)}}, \tag{7.5}$$

where $h = n - 2g$, $t_{h-1}((1-\gamma)/2)$ is the critical point of Student's t-distribution on $h-1$ degrees of freedom, and

$$\text{SSD}_{W_g} = \sum_{W_g}\left(y_i - y_{W_g}\right)^2. \tag{7.6}$$

Here the overall trimming proportion $2\alpha = 2g/n$ is assumed fixed.

The practical statistician, on the other hand, will never be satisfied with a fixed proportion of trimming. If there are no "outliers" and we have a normal sample, one would prefer to use the most efficient estimator and corresponding confidence interval (7.1) which equates to using Eq. (7.5) with $g = 0$ since $\bar{x} = y_{T_0}$ and $s_n^2 = \text{SSD}_{W_0}/(n-1)$. On the other hand, if we have heavy tailed data we would prefer to use $g > 0$ and in the limiting case we would use the median – but the cost is then in the variability of the stability of the estimate of the variability of y_{T_g} and inherently in the length of the confidence intervals. An adaptive approach to this problem was proposed by Tukey and McLaughlin (1963) by putting an upper limit $\tilde{G}(n)$ on the number of observations to trim and then recommending either of the options

(1) choose that $g \le \tilde{G}(n)$ which minimizes $\text{SSW}_g = \text{SSD}_{W_g}/h(h-1)$; or

(2) choose that $g \le \tilde{G}(n)$ which minimizes $t_{h-1}((1-\gamma)/2)\sqrt{\frac{\text{SSD}_{W_g}}{h(h-1)}}$.

Here one is either minimizing an estimate of the variance of the trimmed mean or minimizing the length of the corresponding confidence interval. Jaeckel (1971b) takes up the baton and chooses to allow the trimming proportion α from each tail to vary continuously and to minimize an estimate of the variance of the trimmed mean, $S^2(\alpha)$ say, in his development. Here if $\alpha = 2g/n$, then $S^2(\alpha) = (1 - 1/h)\text{SSW}_g$, whereupon the proposal (1) of Tukey and McLaughlin and of Jaeckel are quite similar. Jaeckel (1971a) restricts α to be positive in the subsequent development of the asymptotic limit theorem for the optimal trimmed mean. An empirical reason for this is noted in Clarke (1994). In particular, if the true distribution is normal, that is the optimal choice for trimming should be alpha equal to zero, the result of employing (1) where we trim up to approximately 50% of the data by choosing $\tilde{G}(n) = \lfloor n/4 \rfloor$ is that the resulting estimated optimal total proportion of trimming $2\hat{\alpha} = 2g/n$ ranges right across the interval $[0, \frac{1}{2}]$ in simulated experiments. Hence, one would tend to trim unnecessarily in this important case. Certainly, the limit theorem which gives an asymptotic distribution equal to the asymptotic distribution of the trimmed mean based at an optimal trimming proportion does not hold since the trimming proportion is not consistent to any value when the data are from the normal distribution. This somewhat disappointing fact subsequently tended to dampen the enthusiasm for using trimmed means to help identify outliers in data in an automatic way from then on. However, the appearance of adaptive estimates in other areas of the literature is now burgeoning since its initial appearance in this form and in the brief summary of Huber (1996, chapter 7). An up-to-date reference based on mainly empirical evidence is that of O'Gorman (2012).

7.1.1 The Functional Form of L-Estimators of Location

A general form for writing *linear combinations of order statistics* or L-estimators is

$$T[F_n] = \sum_{i=1}^{n} a_i y_i; a_i = M\left(\frac{i}{n}\right) - M\left(\frac{i-1}{n}\right).$$

The special case is the (symmetrically) α-trimmed mean that trims a proportion α of observations from each tail, where M is uniform on the subinterval $(\alpha, 1 - \alpha)$. It can be noted that Maronna et al. (2006, p. 31 and p. 348) point out that the exact distribution of the trimmed mean is intractable and that its large-sample approximation is more complicated than for M-estimators and even heuristic derivations are involved. A recent attempt at defining the small sample distribution of the α-trimmed mean is given by Garcia-Pérez (2016).

Another way of writing the estimator is to write

$$T[F] = \int F^{-1}(t)\, dM(t)$$

where for $T[F]$ to be a location functional averaging the quantiles of F we require M to be a distribution on $(0, 1)$ symmetric about one half. For example, the Winsorized mean involves a measure with discrete atoms of probability at α and $1 - \alpha$ in addition to a probability mass of $1 - 2\alpha$ uniformly spread on the interval $(\alpha, 1 - \alpha)$ so that

$$T_W[F] = \int_{\alpha}^{1-\alpha} F^{-1}(t)\, dt + \alpha\left(F^{-1}(\alpha) + F^{-1}(1 - \alpha)\right).$$

When $M(t)$ has a density one can easily write $M(t) = \int_0^t J(u)\, du$, where $J(u)$ can be referred to as a score function. Under some conditions Boos (1979) details

Fréchet differentiability with respect to the Kolmogorov or supremum norm. We detail with permission his Theorem 1 in the following and the conditions under which he attains it.

Suppose the score function J is trimmed near 0 and 1,

$$J(u) = 0 \quad u \in [0, t_1 \cup (t_2, 1],$$ (7.7)

$0 < t_1 < t_2 < 1$. Let F be a fixed distribution function and let $\mathcal{G}$ be the space of all distribution functions.

Theorem 7.1: *Boos (1979)*
If Eq. (7.7) holds and

$$J(u) \text{ is bounded and continuous a.e. Lebesgue and a.e. } F^{-1}$$ (7.8)

then the Fréchet derivative of $T[F] = \int F^{-1}(t) J(t) \, dt$ at F with respect to $d_k \equiv \|\cdot\|_\infty$ is given by

$$T_F'[G - F] = - \int_{-\infty}^{\infty} (G(x) - F(x)) J(F(x)) \, dx .$$

Condition (7.7) rules out the mean functional. As flagged in the introduction of this chapter, the trimmed mean, with a fixed equal positive proportion of observations being trimmed from each tail, is noted by Huber (1977, 1981, 1996), and others including Maronna et al. (2006, formula (10.25)) to have the same influence function as Huber's minimax solution of an M-estimator with $k = F_0^{-1}(1 - \alpha)$ if $F(x) = F_0(x - \mu)$ and F_0 is symmetric about zero. Also the asymptotic distribution is given by

$$\sqrt{n} \left(T[F_n] - \mu \right) \to_d N \left(0, \frac{1}{(1 - 2\alpha)^2} E_F[\psi_k(X - \mu)^2] \right) .$$

The asymptotic variance is the same as that of the Huber M-estimate. Interestingly, the equivalence of Fréchet differentiable estimates to M-estimates is explored in the two publications, Bednarski et al. (1991) and Bednarski and Clarke (1998).

It is pointed out that the trimmed mean is a function of all observations (even those not included in the sum Eq. (7.2)). The effect of outliers, say in one tail, is nullified if the proportion of trimming in each tail is greater than the proportion of outliers, but then "good" observations will be trimmed from the other tail unnecessarily. As we have seen above, the idea of choosing the proportion of trimming adaptively to cast out spurious observations is also stymied when the data are normal, since the adaptive approach here tends to choose positive proportions of trimming when that should otherwise be. A potentially much better approach, that has a principle behind it, is offered in the next section.

Further and more extensive asymptotic theory for L-estimators including for the linear model is detailed in the book Jurečková et al. (2012).

7.2   THE TRIMMED LIKELIHOOD ESTIMATOR

The natural idea of trimming the likelihood before maximizing arises out of considerations of outlier identification and practice in multivariate statistics. Long before any formal writing down of what a trimmed likelihood estimator (TLE) might look like statisticians were regularly identifying unusual observations in multivariate samples by first calculating estimates of normal distribution parameters, then calculating Mahalanobis distances of each observation from those parameters, before making judgments on whether particular observations should be discarded. Maximum likelihood estimates of the data without unusual observations could be used. With the advent of robustness theory, alternative robust estimates were proffered, prior to calculation of Mahalanobis distances from individual observations. See Campbell (1980). The process of identifying unusual observations and re-calibrating estimates could be viewed as a combination of L- and M-estimation. An actual numerical implementation of trimmed likelihood ideas was carried out for the multivariate Gaussian model in Broffitt et al. (1980), though no formal enunciation of the ideas was given. Bednarski and Clarke (1993) motivated by these papers sought to quantify the idea for a general parametric family, having a parametric family of densities $\{f_\theta\}$, using a trimming function J_α akin but not exactly the same as in Boos (1979). The estimator was written down by considering functionals on the product space $\mathcal{G} \times \Theta$ of the form

$$S(F, \theta) = \int \log f_\theta(x) J_\alpha[F\{y : \log f_\theta(y) \le \log f_\theta(x)\}] \mathrm{d}F(x) \qquad (7.9)$$

where

$$J_\alpha(t) = \begin{cases} 0 & \text{if } t \le \alpha \\ 1 & \text{if } \alpha < t \le 1 \end{cases}. \qquad (7.10)$$

The TLE is thus defined as an intuitively appealing estimator as

$$T_\alpha[F] = \arg_\theta \max_\theta S(F, \theta). \qquad (7.11)$$

According to Bednarski and Clarke, for $T_\alpha[F]$, "extremely small values of $\log f_\theta(x)$ are given no weight in the integral expression $S(F, \theta)$. This is equivalent to trimming observations that are least likely to occur as indicated by the likelihood." They termed this idea the *trimmed likelihood principle*. The observations with smallest values of the $\log f_\theta(x)$ are also the values with the smallest values of $f_\theta(x)$ so it is clear one is trimming the likelihood. Bednarski and Clarke go on to describe the asymptotic properties of the TLE functional at the univariate normal distribution location and scale family by examining the functional estimator as a solution of equation

$$L(F, \theta) = \int \phi(x, \theta) J_\alpha[F\{y : \log f_\theta(y) \le \log f_\theta(x)\}] \mathrm{d}F(x) = 0, \qquad (7.12)$$

where $\phi(x, \theta) = (\partial/\partial\theta)\log f_\theta(x)$ and so that the estimating functional is a solution of $L(F, \theta) = 0$. They also describe the differentiability of the estimating functional. The functional derivative is a compact like derivative, and it is shown that the estimating functional for location has the same asymptotic distribution as for the M-estimator known as Huber's skipped mean (Hampel et al., 1986, pp. 158–159). Consequently, the influence function for location is discontinuous at the normal distribution and as a result the functional estimator cannot be Fréchet differentiable at the normal distribution. Another interesting feature of the estimator is that the solution of the estimating equations for scale is not Fisher consistent. Indeed a scaling factor based on the proportion of trimming and the standard normal density is required to make the estimator of scale consistent, even when jointly estimating location and scale. Bednarski and Clarke go on to describe the Fisher consistency of the resulting estimates for location and scale at the normal parametric distribution and give the resulting joint distribution of the asymptotic vector functional estimator using results from Billingsley (1968).

There are other approaches to the trimmed likelihood estimation problem, mainly to do with estimation of location with its natural extension to regression models and then multivariate models assuming a Gaussian distribution. It has arisen in several different forms. In estimation of location, the idea of combining L- and M-estimators appeared in a paper by Rivest (1982), where the estimator (T_n) is defined as a solution of the equation

$$\sum_{i=1}^{n} J_n^* \left(\frac{i}{n+1}\right) \psi \left(\frac{X_{(i)} - t}{s_n^*}\right) = 0 \quad \text{(with respect to } t)$$

where s_n^* is a suitable estimator of the scale parameter and J_n^* is a suitable score function. [A * is used in this book different to Rivest (1982) in order not to confuse symbols already used.] Here Rivest explains how L-estimators are achieved with the choice of $\psi(x) \equiv x$, while M-estimators are the result of a choice of $J_n^*(\cdot) \equiv 1$. The TLE is a special case which links both J^* and ψ. Further asymptotics could be described as suggested in Jurečková et al. (2012, p. 88). Also, in the same year, Butler (1982) studied nonparametric interval and point prediction using data trimmed by a Grubbs-type outlier rule. The principal aim was the identification of a prediction interval for a new observation X_{n+1} from a location model, by choosing to span a fixed proportion $0 < 1 - \alpha < 1$ of the data, where the span of the $100(1 - \alpha)\%$ of connected data supports the smallest sample variance. Butler also details in his Theorem 4.1 the asymptotic distribution of the mean of the data in that span, even when the underlying distribution F may be asymmetric. When the underlying F is chosen or modeled to be normal, the mean of the data in the span corresponds to the TLE of location at the normal distribution, and when $\alpha = \frac{1}{2}$ the estimator is that of the least trimmed squares (LTS) estimator as we shall see in the following.

7.2.1 LTS and Breakdown Point

The LTS estimator was introduced into the literature by Rousseeuw (1983, 1984) along with the *least median of squares* (LMS) estimator in regression modeling. In the same papers, Rousseeuw describes the breakdown point of Hampel (1971) adapted for the finite sample case by Donoho and Huber (1983), arguing that contamination can vary anywhere from zero to approximately 50% of the data; beyond 50% nothing can be learnt. The finite sample breakdown point is the smallest fraction of contaminated data that can replace observations on a given sample that can cause the estimator to take on arbitrarily large aberrant values. We give a definition of breakdown point related to the definitions in Hampel et al. (1986, pp. 97–98) and the summary definition in Rousseeuw and Leroy (1987, pp. 9–10). See Neykov and Mller (2003) for another definition.

Let $\Omega^* = \{X_i \in R;\ i = 1, \ldots, n\}$ be a sample of size n.

Definition 7.1: The breakdown point of an estimator $T_n (X1, \ldots, X_n) \equiv T(\Omega^*)$ is given by

$$\varepsilon^* (T, \Omega^*) = \frac{1}{n}\min\{r : \sup_{\tilde{\Omega}_r} |T(\tilde{\Omega}_r)| = +\infty\},$$

where $\tilde{\Omega}_r$ is any sample obtained from Ω^* by replacing any r of the points in Ω^* by arbitrary values. Subsequently, there is a compact set in which the estimator remains even if we replace any $r - 1$ elements of the sample Ω^* by arbitrary ones. The smallest r/n for which the estimator "breaks down" is the breakdown point.

In the regression setting of Section 3.3, the observations are in fact $X_i \equiv (Y_i, x_{1i}, x_{2i}, \ldots, x_{mi})^T$. The least squares estimator which chooses the regression parameter estimate $T \equiv \hat{\beta}$ to satisfy

$$\sum_{i=1}^{n} \epsilon_i^2 (\beta) = \sum_{i=1}^{n} (Y_i - x_i'\beta)^2 = \min!, \tag{7.13}$$

where $x_i^T = [1, x_{1i}, \ldots, x_{mi}]$ and $\beta = (\beta_o, \beta_1, \ldots, \beta_m)^T$, has in fact a breakdown point of $\frac{1}{n}$ since one outlier is sufficient to carry T over all bounds [cf. Rousseeuw and Leroy (1987, pp. 9–10)].

The LMS estimator satisfies

$$\text{med}_i\left(\epsilon_i^2\left(\boldsymbol{\beta}\right)\right) = \min!, \tag{7.14}$$

and while it is known as a good initial estimator it is also known to have only an asymptotic convergence rate of $n^{-\frac{1}{3}}$. See Davies (1992b) and Andrews et al. (1972). The LTS estimator satisfies, on the other hand,

$$\sum_{i=1}^{h}\left(\epsilon^2\left(\boldsymbol{\beta}\right)\right)_{i:n} = \min!, \tag{7.15}$$

where $\left(\epsilon^2\right)_{1:n} \leq \cdots \leq \left(\epsilon^2\right)_{n:n}$ are the ordered squared residuals. Rousseeuw and Leroy (1987, pp. 124 and 132) choose h to be $\lfloor n/2 \rfloor + \lfloor (p+1)/2 \rfloor$, bearing in mind that there are $p = m + 1$ regression parameters in our definition of the model. Asymptotically, the breakdown point of both LMS and LTS estimators is 0.5. The LTS estimator, on the other hand, converges at a rate of $n^{-\frac{1}{2}}$. Surprisingly, however, at the location model of the standard normal distribution the LTS estimator has a very large asymptotic variance of 14.021 (Rousseeuw and Leroy, 1987, p. 191) and therefore corresponds to an estimator with only 7.3% efficiency. In multivariate estimation, the corresponding estimator to the LTS location estimator is the minimum covariance determinant (MCD) estimator (see Rousseeuw, 1983; Butler et al., 1993). Neykov and Neytchev (1990) in a short communication recognize that the LTS regression estimator can be written as

$$\sum_{i=1}^{h}\left(-\log f_{\boldsymbol{\theta}}\left(x\right)\right)_{i:n} = \min!,$$

which they termed the least trimmed log likelihood estimator for any unimodal density. This was followed up in Vandev and Neykov (1993) who discuss breakdown properties of such an estimator for various values of h. It should be mentioned that trimming the likelihood and then maximizing does not in general, even for unimodal densities $f_{\boldsymbol{\theta}}$, necessarily give consistent estimators of parameters $\boldsymbol{\theta}$. In the case of location of a symmetric distribution, such as the normal, Bednarski and Clarke (1993) establish Fisher consistency of the TLE, and illustrate the Fisher consistency of the "rescaled" (according to a constant to do with the trimming proportion and the underlying model standardized density) solution of the trimmed likelihood equations for scale, and subsequently the Fisher consistency of the joint location and rescaled estimates of scale. They also describe using a functional approach the asymptotic normal distribution of the joint estimator. Earlier descriptions by Rousseeuw (1983, p. 295) using results of Yohai and Maronna (1976) and Butler (1982) do not consider the joint asymptotic distribution of location and scale.

As yet we have not given a description of the computation of the LTS estimator which is a special case of the TLE. For a univariate sample, the algorithm described by Rousseeuw and Leroy (1987, pp. 171–172) can be used to calculate the TLE. Given a general h, not necessarily $h = \lfloor \frac{n}{2} \rfloor + 1$ as for LTS, consider the $n - h + 1$ subsamples

$$\{y_1, \dots, y_h\}, \{y_2, \dots, y_{h+1}\}, \dots, \{y_{n-h+1}, \dots, y_n\}.$$

Each of these subsamples contains h observations, so it is called by Clarke (1994) a (contiguous) $(1 - \alpha)$-shorth for the want of a name. [The shorth is the shortest interval that covers half of the values. The mean of the observations that lie in the shorth was proposed by Andrews et al. (1972) as a robust estimator of location.] For each contiguous $(1 - \alpha)$-shorth one calculates the mean

$$\bar{y}^{(1)} = \frac{1}{h} \sum_{i=1}^{h} y_i, \ \dots, \bar{y}^{(n-h+1)} = \frac{1}{h} \sum_{i=n-h+1}^{n} y_i$$

and the corresponding sums of squares

$$\mathrm{SQ}^{(1)} = \sum_{i=1}^{h} \left\{ y_i - \bar{y}^{(1)} \right\}^2, \ \dots, \mathrm{SQ}^{(n-h+1)} = \sum_{i=n-h+1}^{n} \left\{ y_i - \bar{y}^{(n-h+1)} \right\}^2$$

and then the LTS solution corresponds to the mean $\bar{y}^{(j)}$ with the smallest associated sum of squares $\mathrm{SQ}^{(j)}$. In extensive simulations in Clarke (1994), it was always the case that the pair $\left(\bar{y}^{(j)}, \sqrt{\mathrm{SQ}^{(j)}/h} \right)$ for which $\mathrm{SQ}^{(j)}$ is the smallest were also solutions of the location and scale trimmed likelihood equations (7.12) as confirmed by numerical algorithms. Subsequently, we denote these as $\left(T_\alpha[F_n], \bar{\sigma}_\alpha[F_n] \right)$. Also it can be noted that the mean of the observations in the contiguous $(1 - \alpha)$-shorth for which $\mathrm{SQ}^{(j)}$ is a minimum is the mean of the span for which the variance is minimized in the nomenclature of Butler (1982).

7.2.2   TLE Asymptotics for the Normal Distribution

Since the normal distribution is the focus of much of the effort, we as in Bednarski and Clarke (1993) write down the Fisher consistency result of Lemma 2.3 of that paper here.

Lemma 7.1: *Bednarski and Clarke (1993)*
Let $F_{\mu,\sigma}$ denote the normal distribution with unknown mean μ and scale σ. Let $\theta = (\mu, \sigma)^T$ and $\bar{\theta}^T = (\bar{\mu}, \bar{\sigma})^T \equiv \left(T_\alpha[F_\theta], \bar{\sigma}_\alpha[F_\theta] \right)^T$ be a solution of (7.12). Then

$$\bar{\mu} = \mu \tag{7.16}$$

and

$$\frac{(1-\alpha)\,\overline{\sigma}^2}{1-\alpha-\sqrt{2/\pi}z_{\alpha/2}\exp\left(-\frac{1}{2}z_{\alpha/2}^2\right)} = \sigma^2. \tag{7.17}$$

The proof given in the paper uses the definition of the trimmed likelihood equations after considering first the equation for location with scale fixed and then the equation for scale with location fixed, and then the proof with both location and scale as variables in the solution for the Eq. (7.12). That is, $L\left(F_\theta, \overline{\theta}\right) = 0$. Lemma 7.1, thus, gives a rescaled solution of the variance that is formed from the trimmed likelihood estimate of scale, in order that one gains a Fisher consistent estimate for σ^2.

As has been indicated, there are a variety of ways of obtaining the asymptotic distribution of the TLE of location modeled at the normal distribution. As with the bulk of robustness studies, we can construct the estimator using a parametric family $\{f_\theta\}$ being the normal but examine the distribution of the estimator under a more general distribution F. The following specific result can be arrived at via Theorem 4.1 of Butler (1982), Theorem 4 and Remark of Rousseeuw and Leroy (1987, pp. 180–181), or results of Theorem 3.2 and subsequent Remark 3.2 of Bednarski and Clarke (1993). It reiterates part of the summary in the introduction of Bednarski and Clarke (2002, p. 1), though here to conform to previous notation in this chapter we refer to the true center of symmetry as μ rather than μ_o that was used in papers by Bednarski and Clarke.

Theorem 7.2: *Consider $0 < \alpha < 0.5$ fixed. Let $T_\alpha[\cdot]$ be the location estimator constructed from the TLE assuming a normal parametric family (either location or location and scale). See Eq. (7.11) or Eq. (7.12). Let $F_n(x)$ be the empirical distribution function formed from the sample $X_1, \ldots, X_n$ generated by the symmetric distribution $F(x) = F_o(x - \mu)$, where μ is the center of symmetry. Also let F have a continuous positive density f at $\mu + x_\alpha$, where $x_\alpha = F_o^{-1}(1 - \alpha/2)$. Then one has*

$$\sqrt{n}\left(T_\alpha[F_n] - \mu\right) \to_d N\left(0, V(\alpha, F)\right)$$

where

$$V(\alpha, F) = \frac{\int_{\mu-x_\alpha}^{\mu+x_\alpha}(x-\mu)^2\,dF(x)}{\left\{1-\alpha-2x_\alpha f\left(x_\alpha+\mu\right)\right\}^2}$$

$$= \frac{\int_{-x_\alpha}^{x_\alpha} x^2\,dF_0(x)}{\left\{1-\alpha-2x_\alpha f_0\left(x_\alpha\right)\right\}^2}. \tag{7.18}$$

Corollary 7.1: In the special case of $F(x) = \Phi\{\frac{x-\mu}{\sigma}\}$ then

$$\sqrt{n}\left(T_\alpha[F_n] - \mu\right) \to_d N\left(\cfrac{0, \sigma^2}{\left\{1 - \alpha - \sqrt{2/\pi}z_{\alpha/2}\exp\left(-\tfrac{1}{2}z_{\alpha/2}^2\right)\right\}}\right).$$

□

From the above corollary, the asymptotic variance in the special case that the underlying distribution F is the normal distribution is minimized when α is zero. This is what we know from standard statistical inference, since the most efficient estimator for location according to the full Cramér–Rao theory is in fact the sample mean. The sample mean is the TLE for location when $\alpha = 0$, that is the full likelihood estimate. On the other hand, if the underlying distribution was in fact heavy tailed, it would inherently be better to trim than not to trim, as our experience and that of previous authors going right back to Tukey (1960) would suggest. Tukey's study, showed for example, that the 5% trimmed mean performed reasonably well at the normal distribution and retained high efficiency in small departures from the normal involving a mixture distribution with an increasing proportion of contamination from a component distribution with the same center of symmetry but nine times the variance. Compare, for example, the discussion in Chapter 1 around Figure 1.9. In a limited simulation study, Bednarski and Clarke (1993, section 5, table 1) show that for sample sizes as low as 20, the relative agreement between sample standard errors and standard errors is based on the asymptotic distribution for the TLE of location. Also while there is a small loss of efficiency for the same amount of overall trimming as the trimmed mean at the normal model, this is somewhat compensated for as demonstrated in their table 2 by a significant reduction in estimated bias when contamination is allowed to be asymmetric. An example is when the underlying distribution is $F(x) = 0.95\Phi\{x\} + 0.05\Phi\{\frac{x-3}{0.1}\}$. In 1000 replications of a sample of size 100 and assuming a fixed proportion $\alpha = 0.1$ of overall trimming, the estimated bias of the TLE was 0.0135 compared to that of estimated bias of the trimmed mean of 0.1168, when the estimated standard errors of the two estimates were comparable at 0.1165 and 0.1010 respectively. Here the trimmed mean is chosen to trim 5% of observations from each tail. It could be argued that one compare estimators with the same overall breakdown point. It is shown, for example, in Clarke (1994, Proposition 2.1) using an extension of the argument of Rousseeuw and Leroy (1987, Theorem 6, p. 184) that for $0 \leq \alpha < \frac{1}{2}$ where $g = n\alpha$ is an integer and for any univariate sample, call it Ω^*, the breakdown point of the trimmed likelihood estimator is $\varepsilon^*\left(T_\alpha, \Omega^*\right) = (g+1)/n$ which converges to α as $n \to \infty$. For the same overall proportion of trimming, the asymptotic limit of the breakdown point of the trimmed mean is $\alpha/2$. The TLE obviously can account for a full proportion $\alpha < \frac{1}{2}$ of contamination that is potentially on one tail of the model distribution, whereas the trimmed mean can only account for a proportion $\alpha/2$ from one tail before it begins to breakdown. Clearly, if one is suspicious of asymmetric contamination it may be preferable to use the TLE and if one has an idea of the proportion of contamination then the recommendation is to choose a trimming proportion in excess of that.

The reality is that according to Hampel et al. (1986, section 1.2c) the frequency of gross errors in routine data is between 1 and 10%. With high-quality data the proportion may be even smaller. Seemingly then in typical cases we should, if we are to act accordingly, trim a greater percentage of data, perhaps 20% or even a little more. But this does not guard us against those instances where the contamination can be up to approximately 50%, the argument upon which most of the book by Rousseeuw and Leroy (1987) is based. For example, trimming 50% of the data whereupon one is using LTS those authors recommend in the light of the poor efficiency of that estimator to choose, after implementing LTS, a one-step improvement which could be using the Tukey bisquare M-estimator with a single Newton iteration. See Bickel (1975) for a discussion of this in the case of the linear model and how it retains the efficiency of the fully iterated estimator at the model. In estimation of multivariate location, Davies (1992a) on the other hand shows that a two-step M-estimator is Fréchet differentiable even starting from the minimum volume ellipsoid estimator which in the univariate case is the LMS estimator, the latter of which is not even root n consistent.

The fact is that if we choose a fixed proportion of trimming for the TLE for the normal location problem, the estimator will be unstable, particularly in terms of a contamination about the critical point $z_{\alpha/2}$, as is the case also for Huber's skipped mean. This is where the discontinuity in the influence function is found and which precludes Fréchet differentiability of the estimator which is a necessity for infinitesimal robustness. Choosing to trim excessively in the neighborhood of 50% trimming exposes one to poor efficiency. The suggestion of Davies is an asymptotic one, yielding highly efficient estimates that are also robust, but when n is so large that we can ignore the difference in distribution between the two-step M-estimator and the infinite step M-estimator, one can question why one needs efficiency, albeit the Newton algorithm converges fast. For smaller samples and also more complex parameter problems, there is no guarantee that for all samples that the LTS estimator will be in the numerical domain of attraction of the consistent Fréchet differentiable root of the M-estimating equation. Two Newton-steps could take the estimator further away to another root. Rousseeuw and Leroy's solution of one-step, on the other hand, is not guaranteed to be Fréchet differentiable. The problem is that we usually do not know the proportion of contamination. It could be from anywhere between zero and almost 50%. While the next section outlines an adaptive procedure or procedures to align ourselves with this possibility, we will eventually return to the question of estimation and the use of Fréchet differentiable statistical procedures. One should be careful to distinguish between the objectives, one of obtaining a highly robust and reasonably accurate statistical estimate and one of identifying gross errors in the data. The two are essentially different objectives, though they fall in the same domain of robust discussion.

7.3 ADAPTIVE TRIMMED LIKELIHOOD AND IDENTIFICATION OF OUTLIERS

A natural suggestion is to minimize the asymptotic variance, and since we do not know F, we follow the line of Jaeckel's approach and minimize an estimate of the asymptotic variance, though in this section we are using a trimmed likelihood estimate rather than a trimmed mean, unlike Jaeckel. There are two main possibilities, but not all here. We can either minimize an estimate of $V(\alpha, F)$ based on

$$v(\alpha, F_n) = \frac{\overline{\sigma}_\alpha^2[F_n]}{\left\{1 - \alpha - \sqrt{2/\pi}z_{\alpha/2}\exp\left(-\frac{1}{2}z_{\alpha/2}^2\right)\right\}^2}, \tag{7.19}$$

or, basing the estimate on the large sample consistent estimate of $V(\alpha, F)$ minimize instead

$$v^*(\alpha, F_n) = \frac{(1-\alpha)\overline{\sigma}_\alpha^2[F_n]}{\left\{1 - \alpha - \sqrt{2/\pi}z_{\alpha/2}\exp\left(-\frac{1}{2}z_{\alpha/2}^2\right)\right\}^2}. \tag{7.20}$$

In empirical investigation, both of these options have been investigated, the most seemingly successful option in initial studies of finite samples being reported in Clarke (1994) as that of option 4 in that paper, which is equivalent to minimizing $v(\alpha, F_n)$. On a finite sample basis, one can view the minimization as follows. If there are a proportion α of genuine extreme outliers in the sample then trimming that proportion of outliers, and then calculating $\overline{\sigma}_\alpha^2$ should give a good estimate of σ^2. If one trims less than a proportion α of outliers, the resulting $\overline{\sigma}_\alpha^2$ would be an estimate of variance including some outliers and the result would involve the numerator swamping the denominator in the expression in Eq. (7.19). But generally increasing the proportion of trimming increases the effect of the denominator again in that expression. Since this is an investigation into outliers, we trim discrete proportions $\alpha = g/n$ where g ranges over the values, $g = 0, 1, 2, \ldots, G^*(n)$, where $G^*(n)$ is either an upper bound on the number of suspected outliers, a bound enforced by the computational complexity of computing, or a value chosen to give a maximum breakdown point of the overall resulting estimator. Of course, in estimation of location there are few examples with bounds on the computing, but see Clarke (2008, Example 7.2) for an example in regression where the computing soon becomes out of bounds. In considering breakdown bounds of a resulting adaptive location estimator, Clarke (1994, Proposition 2.2) indicates the breakdown bound for the estimator of option 4 is $(G^*(n) + 1)/n$ whenever $G^*(n)/n < 1/2$, whereupon trimming up to approximately one half the observations achieves the maximum breakdown point. Hence, we have the adaptive trimming proposal to choose to trim a proportion of observations $\hat{\alpha} = \tilde{g}/n$ so that

$$v\left(\hat{\alpha}, F_n\right) = \min_{0 \le g \le G^*(n)} v\left(\frac{g}{n}, F_n\right). \tag{7.21}$$

As an illustration of how this works, we consider the following example illustration from section 5.3 of Clarke (1994) with some embellishments. The full data set with a classical outlier analysis encompassing weights is discussed in Clarke and Lewis (1998). A related discussion of the problem can also be found in Clarke (1997).

Example 7.1: In a data set from the Western Australian goldfields, company 1 used its mill to crush and treat ore that was purchased from company 2. Payment was on the basis of recoverable gold content less 2 grams per ton (g/t) treatment charge. We list the full set of gold assays (one observation was omitted from the list in Clarke (1994) even though the analysis was done on the full 48 observations given here.) Assuming the data were independent identically distributed observations, the problem is to determine the "best" estimate of mean recoverable gold content. In doing so one outlier is identified.

14.67, 13.73, 20.25, 11.52, 10.94, 10.39, 15.71, 11.43, 11.46, 13.05, 11.97, 12.96, 9.00, 12.78, 17.40, 11.17, 18.00, 7.52, 12.74, 10.18, 13.96, 10.94, 12.40, 8.15, 17.81, 5.04, 25.50, 8.72, 10.00, 6.75, 10.02, 15.53, 61.50, 9.76, 14.04, 5.67, 9.43, 13.05, 10.80, 24.45, 8.18, 6.84, 8.35, 5.00, 16.10, 10.94, 16.76, 12.94

The raw histogram throws up one large outlier and both the histogram and the Q–Q plot suggest that the data are skew (Figure 7.1). As with typical gold assay data (Clarke, 1997, and the references therein), taking logarithms of the data reveal a histogram that apart from the obvious outlier suggest a normal distribution. Here the adaptive TLE chooses $\tilde{g} = 1$. The span which reveals the smallest sum of squares in the trimmed likelihood estimate of the logged data is in fact $\left(y_1, \ldots, y_{47}\right)$, the 48th and largest observation y_{48} being trimmed. This is in fact $\log 61.5$ which is the logarithm of the 33rd observation in the original sample. The original logged gold assay sample reveals a mean of $\bar{x} = 2.475$ which on exponentiating gives the geometric mean $\bar{Y} = 11.88$. The 95% confidence interval for the mean of the logged gold assays is $(2.35, 2.60)$. On the other hand, if we believe that observation 33 is truly a gross error and not to be included, then the logged gold assay sample reveals a mean of $\bar{x} = 2.44$ and on exponentiating gives a geometric mean $\bar{Y} = 11.47$, while the 95% confidence interval for the mean of the logged gold values without the 33rd observation is $(2.33, 2.55)$. Of course, if we are to adaptively trim routinely, it is clear that some samples would be trimmed even if the logged data were generated from a normal distribution. In fact in simulations very few samples are trimmed in this sample size range using option 4. For example, from table 1 in Clarke (1994) for samples of size $n = 40$ only 44 samples in 10000 generated from a normal distribution involved trimming any positive number of observations, and the average overall trimming proportion was 0.0001. So the effect of trimming adaptively is going to have little deleterious effect on confidence intervals. Nevertheless, some adjustment

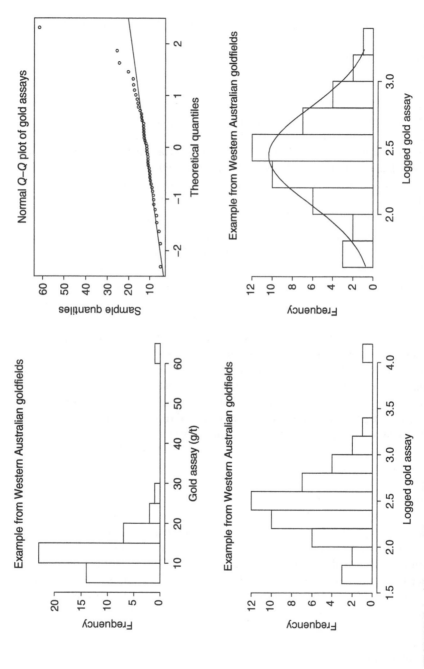

FIGURE 7.1 Histogram of raw gold assay and Q–Q plot of raw gold assay data followed by histogram of full set of logged gold assay data and then a histogram of logged gold assay data with fitted line plot of normal density for data with outlier removed.

for trimming in resultant confidence intervals could be made as was tentatively attempted in procedure D given in Clarke (1994).

One has to be aware of the fact that the geometric mean is estimating e^μ which in fact is the median of the log normal distribution, the hypothesized distribution of the original data. For instance, μ is the mean of the hypothesized normal distribution of the logged data. It is also the median of that distribution. Exponentiating preserves order in the resultant original distribution and thus we are estimating the median of the log normal distribution by using $\overline{Y}$. The actual mean of the hypothesized distribution of the original data is $e^{\mu+\frac{\sigma^2}{2}}$. Here σ^2 is the variance of the hypothesized normal distribution of the logged data. Taking account of this, one gains estimates of the mean of the original data of 13.04 and of the data with the 33rd observation removed of 12.24. The difference in estimates is a substantial fraction of the treatment charge. $\qquad\qquad\qquad\qquad\qquad\qquad\qquad\qquad\qquad\quad$ □

Small samples do not appear either to be a problem for the method to identify gross errors. The simulations in Clarke (1994) illustrate this, and the celebrated example of the Cushny and Peebles (1905) on prolongation of sleep involving the 10 pairwise differences (i.e. the set of differences between drug effects per subject) ordered as follows:

$$0.0, 0.8, 1.0, 1.2, 1.3, 1.3, 1.4, 1.8, 2.4, 4.6$$

again identifies $\tilde{g} = 1$ with the one observation being trimmed of 4.6. This observation is identified as an outlier in numerous studies with examples being Hampel et al. (1986, pp. 78–80) and Staudte and Sheather (1990, pp. 97–98).

Of course, the method can also identify more than one gross error simultaneously. As an important example, Clarke (1994) also gives the results of analyzing the celebrated Newcomb's measurements of the passage time of light recorded in Stigler (1977, table 5). This gives 66 measurements which were the third series of measurements recorded by Newcomb in 1882. Two suspicious values recorded as the only negative numbers in the table are identified as being adaptively trimmed by option 4, since the value $\tilde{g} = 2$ corresponding to an $\hat{\alpha} = 2/66$ that minimizes Eq. (7.21) is such that the span of the positive ordered observations $\left(y_3, y_4, \dots, y_{66}\right)$ provides a minimum of the quantity Eq. (7.19). A recent discussion of Ghosh and Basu (2017) goes to great lengths to illustrate their methodology indeed down weights these suspicious points.

7.4 ADAPTIVE TRIMMED LIKELIHOOD IN REGRESSION

In this section we begin by recapping the introductory discussion of the adaptive trimmed likelihood algorithm (ATLA) relayed in Clarke (2000c). The methodology was also illustrated briefly with an example in Clarke (2008, section 7.2.1).

Again we refer to the regression model of Section 3.3 in the current book and adapt the notation of this discussion accordingly. There are some differences between the calculation of an adaptive TLE for location and for regression. These include considerations of breakdown bound, asymptotic variance, and ease or otherwise of computation.

The maximum asymptotic breakdown bound for location of the TLE is α which is maximized for $\alpha = 1/2$ which is when the estimator at the normal distribution is known as LTS. The finite sample breakdown bound is then $\varepsilon\left(T_{1/2}, \Omega^*\right) = \frac{\lfloor (n+1)/2 \rfloor}{n}$ (Rousseeuw and Leroy, 1987, Theorem 6, p. 184). If n is even, the breakdown bound of LTS is one half and nothing can be decided if $n/2$ points are chosen arbitrarily. Hence in any adaptive framework, we should choose $G^*(n) \leq \lfloor n/2 \rfloor$ to be the maximum number of points to trim in the adaptive trimmed likelihood. If $G^*(n) = \lfloor n/2 \rfloor$ we get by Clarke (1994, Proposition 2.2) that the breakdown bound of the adaptive TLE is asymptotically exactly one half regardless of whether n is even or odd. By the proof of Proposition 2.1 and the proof of Proposition 2.2 of that paper choosing to replace $G^*(n) + 1$ points arbitrarily can cause the estimator to break down. For n odd that is not the case when $G^*(n)$ or less points are chosen to be replaced arbitrarily. For n even it is not the case when $G^*(n) = n/2$ that choosing $n/2 - 1$ points or less and assigning them arbitrarily can break down the estimator. However, choosing $n/2$ points and assigning them arbitrarily can break down the estimator.

For regression the intention is that there is a model

$$Y_j = x_j^T \beta + \epsilon_j; j = 1, \ldots, n, \tag{7.22}$$

where x_j and β are $p \times 1$ vectors of covariates and regression coefficients. The assumption on what much inference is based is that the ϵ_j are independent random variables which are $N\left(0, \sigma^2\right)$ say. As mentioned above in implementing LTS, Rousseeuw and Leroy (1987) choose the value $h = \lfloor n/2 \rfloor + \lfloor (p+1)/2 \rfloor$ to maximize the breakdown point. In analogy with the univariate location case, it is then appropriate to choose any adaptive estimator based on trimming anywhere from zero to

$$G^*(n) = n - \lfloor n/2 \rfloor - \lfloor (p+1)/2 \rfloor \tag{7.23}$$

observations where one minimizes a measure of variance. Since there is no longer a natural $(1 - \alpha)$-shorth or span from which an algorithm can proceed to compute the regression estimator, if one wants to trim exactly $g = n - h$ observations in calculating the corresponding TLE, one is forced to calculate the objective function in Eq. (7.15) by choosing each set of possible $h = n - g$ observations and calculating the residuals of the corresponding least squares estimator (having removed already the corresponding set of g potential observations to trim). The set of h observations giving the minimum value points to the corresponding set of g observations to trim. The TLE of β is then the least squares estimator using the untrimmed set of observations. It can be immediately observed that this is a potentially computationally intensive procedure, requiring one for any fixed g to examine $N = \binom{n}{g}$ sets of least squares estimates. This is even more so if indeed one wishes to check a further objective variance in the adaptive formulation where one trims anywhere from g equal to zero up to $G^*(n)$ observations, since there is not necessarily a recursive formula.

This is the cost one makes for wishing to examine all possible sets of observations as potential outliers via adaptive trimmed likelihood methodology. The trimmed likelihood estimate, the trimmed observations, and the trimmed sum of squared errors in Eq. (7.15) for each g first needs to be recovered. We repeat here the algorithm from Clarke (2000c) for calculating a trimmed likelihood estimate. This algorithm does not include what is called a random sampling procedure discussed, for example, in Rousseeuw and Leroy (1987, chapter 5) used to calculate LTS, for such a procedure may include potential outliers albeit with small probability.

TLA:

1. Initialize $S_{0\alpha}^2 = \infty$, $l = 1$ and $\tilde{J} = \{1, \dots, g\}$
   REPEAT WHILE $l \leq N$
2. Let $\boldsymbol{\beta}_l$ solve $Y_i = \boldsymbol{x}_i^T \boldsymbol{\beta}_l$ for $i \notin J_l$
3. Let $S_l^2 = \sum_{i=1}^{h} \left(\epsilon^2 (\boldsymbol{\beta}_l) \right)_{i:n}$
4. If $S_l^2 < S_{0\alpha}^2$ then do:
   (i) $\tilde{\boldsymbol{\beta}} \leftarrow \boldsymbol{\beta}_l$
   (ii) $S_{0\alpha}^2 \leftarrow S_l^2$
   (iii) $\tilde{J} \leftarrow J_l$
5. $l \leftarrow l + 1$

On running the above algorithm for fixed g then define $\hat{\boldsymbol{\beta}}(g) = \tilde{\boldsymbol{\beta}}$, $\tilde{\sigma}^2(g) = S_{0\alpha}^2 / (h - p)$ and the set of observation numbers, $\tilde{J}$, is effectively a function identifying g potential outliers; call it $\tilde{J}(g)$.

To give the ATLA estimator, in analogy with Eqs. 7.18 and 7.20 for the univariate problem, define

$$\text{var}\left(\sigma^2, \alpha\right) = \frac{\sigma^2}{\left\{ 1 - \alpha - \sqrt{\frac{2}{\pi}} z_{\alpha/2} \exp\left(-\frac{1}{2} z_{\alpha/2}^2\right) \right\}^2}, \tag{7.24}$$

and let

$$V_n(g) = \text{var}\left(\tilde{\sigma}^2(g), \frac{g}{n}\right), \tag{7.25}$$

and choose $\tilde{g}$ to satisfy

$$V_n(\tilde{g}) = \min_{0 \leq g \leq G^*(n)} V_n(g). \tag{7.26}$$

Then

$$\hat{\beta}_{\text{ATL}} = \hat{\beta}(\tilde{g}) \quad \text{and} \quad \hat{\sigma}^2_{\text{ATL}} = \hat{\sigma}^2(\tilde{g}).$$

The choice for $G^*(n)$ is Eq. (7.23) or something smaller chosen by the statistician. The outliers identified by ATLA are then given by $\tilde{J}(\tilde{g})$.

The divisor $h - p$ used in the definition of $\tilde{\sigma}^2(g)$ above gives the usual least squares analysis of variance estimate of σ^2 when there are no outliers.

Example 7.2: *Modified Data of Wood-Specific Gravity*
Amongst other data sets Rousseeuw (1984) motivated LMS with an example containing multidimensional real data (Table 7.1). Table 2 of that paper gives 20 observations for which there are five independent variables and an intercept. The data were formulated by replacing four observations from a data set given by Draper and Smith (1966, p. 227) by outliers. Observations 4, 6, 8, and 19 are identified using LMS and a scan of the LMS residuals as being outlying. These observations do not appear to be obvious outliers from the least squares analysis. Table 7.2 demonstrates how ATLA picked out the four outliers. Here Eq. (7.23) gives $G^*(20) = 7$. When $g = 7$ there are N = 77520 least squares estimates to be calculated to evaluate $\hat{\beta}(7)$. This took a MATLAB algorithm written to implement ATLA 1735 cpu seconds to run on a Sun Sparcstation. With $\tilde{g} = 4$ exactly those four observations previously identified above are given by $\tilde{J}(\tilde{g})$. The resulting ATLA estimator corresponds to the least squares estimator with these observations deleted from the sample, thus giving the more efficient estimated regression equation $\hat{y} = x^T\hat{\beta}_{\text{ATL}}$ given by

$$\hat{y} = 0.2174x_1 - 0.0850x_2 - 0.5643x_3 - 0.4003x_4 + 0.6074x_5 + 0.3773,$$

where $x_1, \ldots, x_5$ represent the five independent variables. It is more efficient than LTS or LMS. The LTS estimator is obtained as the least squares estimator after discarding seven observations from the sample of 20.   □

It can be seen from Example 7.2 that there can be no recursive procedure to select the outliers since while $\tilde{J}(1) \subset \tilde{J}(2)$ it is not the case that $\tilde{J}(2) \subset \tilde{J}(3)$ and the final outliers settled upon are $\tilde{J}(4)$ which in fact has no overlap at all with the previous three sets in $\tilde{J}(1) \cup \tilde{J}(2) \cup \tilde{J}(3)$. Clarke (2000c) examines three other data sets found in the literature. These are known as the telephone data, the stack loss data, and the Scottish hill races data, all for which $\tilde{J}(g) \supset \tilde{J}(g-1)$ for $g = 1, \ldots, G^*(n)$ for values of $G^*(n)$ that could allow ATLA to be feasibly computed. Recursive algorithms to calculate successive values of $V_n(g)$ would thus

TABLE 7.1   Modified Data on Wood-Specific Gravity

Index	x_1	x_2	x_3	x_4	x_5	y
1	0.573	0.1059	0.465	0.538	0.841	0.534
2	0.651	0.1356	0.527	0.545	0.887	0.535
3	0.606	0.1273	0.494	0.521	0.920	0.570
4	**0.437**	**0.1591**	**0.446**	**0.423**	**0.992**	**0.450**
5	0.547	0.1135	0.531	0.519	0.915	0.548
6	**0.444**	**0.1628**	**0.429**	**0.411**	**0.984**	**0.431**
7	0.489	0.1231	0.562	0.455	0.824	0.481
8	**0.413**	**0.1673**	**0.418**	**0.430**	**0.978**	**0.423**
9	0.536	0.1182	0.592	0.464	0.854	0.475
10	0.685	0.1564	0.631	0.564	0.914	0.486
11	0.664	0.1588	0.506	0.481	0.867	0.554
12	0.703	0.1335	0.519	0.484	0.812	0.519
13	0.653	0.1395	0.625	0.519	0.892	0.492
14	0.586	0.1114	0.505	0.565	0.889	0.517
15	0.534	0.1143	0.521	0.570	0.889	0.502
16	0.523	0.1320	0.505	0.612	0.919	0.508
17	0.580	0.1249	0.546	0.608	0.954	0.520
18	0.448	0.1028	0.522	0.534	0.918	0.506
19	**0.417**	**0.1687**	**0.405**	**0.415**	**0.981**	**0.401**
20	0.528	0.1057	0.424	0.566	0.909	0.568

Source: Rousseeuw (1984). Reproduced with the permission of Taylor & Francis.

TABLE 7.2   ATLA Analysis on Modified Data on Wood-Specific Gravity Including a Constant

	g	$V_n(g) \times 10^4$	$\tilde{J}(g)$
	0	5.8	—
	1	7.1	11
	2	9.0	3,11
	3	10.7	7,11,14
$\tilde{g} =$	**4**	**4.5**	**4,6,8,19**
	5	4.7	4,5,6,8,19
	6	5.9	4,5,6,8,12,19
	7	5.9	1,4,5,6,7,8,19

Values in bold are those achieved when $V_n(g)$ is a minimum.

have worked if we had known in advance this was going to be the case, which inevitably we do not.

The telephone data given in Rousseeuw and Yohai (1984) consists of 24 observations from 1950 to 1973 with the dependent variable being the total number of international calls made, and the independent variable is the year. The data contains heavy contamination and observations 14–21 or years 1963–1970 are spurious, since in fact according to Rousseeuw and Yohai, in this period another recording system was used, which only gave the total number of minutes of these calls. The value in Eq. (7.23) is $G^* (24) = 11$. The ATLA estimator corresponds to calculating the least squares estimator on the data set with observations in $\tilde{J}(8)$ deleted so there are eight observations identified as outlying. Here $\tilde{J}(8)$ corresponds exactly to observations 14–21.

The ATLA analysis of the stack loss data of Brownlee (1965, p. 454) illustrates that the algorithm should not be applied routinely in all situations. The data are 21 observations on losses of ammonia from an oxidation plant and there are three independent variables and a constant. Several authors arrive at the conclusion that there are four suspected outliers, corresponding to observations 1, 3, 4, and 21. Curiously, in the ATLA investigation of the data it was observed that as for the telephone data $\tilde{J}(g) \supset \tilde{J}(g-1)$ for $g = 1, 8$. Since $G^* (21) = 9$ in Formula (7.23) proved out of bounds for computing $\hat{\beta}(9)$ because of computational time needed, it was assumed that $\tilde{J}(9) \supset \tilde{J}(8)$, whence an obvious adjustment to the algorithm allowed easy computation of $\hat{\beta}(9)$ under that assumption. Since the publication of Clarke (2000c) increase in computing power allowed easy computation of $\tilde{J}(9)$ and confirmed that $\tilde{J}(9) \supset \tilde{J}(8)$. Importantly, the first four observations given by the algorithm as potential outliers were observations 21, 4, 3, and 1 in fact coincided with $\tilde{J}(4)$. However, routine running of the algorithm ATLA yielded $\tilde{g} = 0$, giving zero observations trimmed, whereupon ATLA and least squares agree. The objective function minimized has another local minima $V_{21}(4)$ not much different to $V_{21}(0)$. It is observed by Atkinson (1981, 1986a) and Chambers and Heathcote (1981) that the data are not normally distributed and should be transformed.

Atkinson (1986b) originally reported the Scottish hill races data and analyzed that data set. The dependent variable is the record time in minutes and the independent variables are distance in miles and climb feet. There are 35 observations whence a bound on computation time of ATLA required $G^* (35) = 8$, whereas formula (7.23) yields $G^* (35) = 16$. Atkinson's analysis illustrates how normal plots of studentized residuals reveal observations 7 and 18 as outliers. When these two observations are deleted then an additional plot reveals that observation 33 is an outlier, the point being that observations 7 and 18 mask the presence of an outlier in observation 33. LMS in fact identifies observations 7, 18, 11, 33, and 35 as having the largest absolute residuals and worthy of further attention, but Atkinson finds observations 11 and 35 agree with the bulk of the data. ATLA identified $\tilde{g} = 3$ and $\tilde{J}(3) = \{18, 7, 33\}$ is exactly the set of observations identified by Atkinson.

Despite the computational intensity, an encouraging study of the empirical performance of ATLA was carried out in Clarke (2000c) involving both size and power in detecting whether or not there were outliers. The study included potential points for outliers and also good points which could be high leverage (Clarke, 2008, Chapter 7) or not. These simulations with samples of size $n = 15$ were carried out for simple linear regressions in the ilk of Kianifard and Swallow (1989, 1990) and also Hadi and Simonoff (1993). The study of Hadi and Simonoff advises choosing significance levels, for example implicitly say 5% although 1%

is possible. Unlike that study the size of the ATLA procedure is less than 1% in simulations but with remarkably high power.

When n is small to medium one identifies the outliers and hopefully only the outliers, and then retains all the other observations for use in the estimation, leading to an efficient and robust estimator, not throwing almost one half the data away as appears to be the case with LTS.

7.5   WHAT TO DO IF n IS LARGE?

The modus operandi of the previous section is to identify outliers subject to the normal model of errors. In many experiments these outliers may be of interest themselves and one wants to identify such to investigate the peculiarities of the experiment. The method is hampered, however, by the unreasonable computing time that may be expected to be used when the sample size, n, becomes large. A possibility is to use first a high breakdown point estimator and then apply the location only adaptive procedure to the residuals from that high breakdown point estimator. For small-to-medium sample sizes, this procedure is not necessarily going to work, since high breakdown methods are based on asymptotics and the argument for small-to-medium n would be that the residuals are correlated as, for example, the least squares residuals. Carroll and Ruppert (1988) make the point that robust estimators often provide regression diagnostics as by-products and these can be quite useful, but they are not substitutes for specifically designed diagnostics, for example, as through least squares regression diagnostics. Staudte and Sheather (1990, p. 277) also rightly point out that, unfortunately, when there are several influential points some of those points can exert a strong influence on the least squares fit that other influential points can be masked and hence not detected by least squares-based diagnostics. The effects of masking are overcome in ATLA in the previous section in small-to-medium samples. For large n we can expect high breakdown procedures to perform as according to their asymptotics, for which they are designed.

A useful estimator is the MM-estimator for which we adopt the paragraph of discussion in Clarke (2008, p. 170). In a line of research (Yohai, 1987; Yohai and Zamar, 1988; Yohai et al., 1991), MM-estimators were chosen to gain high efficiency and high breakdown point. MM-estimators are based on two loss functions, ρ_0 and ρ_1, and the regression MM-estimator $\hat{\beta}_n$ satisfies the equation

$$\frac{1}{n} \sum_{i=1}^{n} \rho_1' \left(\frac{Y_i - x_i^T \beta}{\hat{\sigma}_n} \right) x_i = 0,$$

where $\hat{\sigma}_n$ is a scale S-estimator (Rousseeuw and Yohai, 1984). That is, $\hat{\sigma}_n$ minimizes the M-scale $\hat{\sigma}_n(\beta)$, defined implicitly by the equation

$$\frac{1}{n} \sum_{i=1}^{n} \rho_0 \left(\frac{Y_i - x_i^T \beta}{\hat{\sigma}(\beta)} \right) = b.$$

These estimates are supported in statistical computing packages such as S-Plus and R.

Asymptotics for these estimators have been discussed in Omelka and Salibián-Barrera (2010) and Fasano et al. (2012) (See also Salibián-Barrera and Zamar (2004).) The results of Fasano et al. prove weak continuity of MM-estimators and a type of differentiability sufficient to prove asymptotic normality under some complicated conditions that include ρ_0 and ρ_1 being bounded and twice continuously differentiable. These underscore the difficulty in this quite complex estimation problem. Perhaps the more satisfying results are those of Omelka and Salibián-Barrera who have some success attacking the problem of uniform convergence of MM-estimators of regression in contamination neighborhoods of the underlying distribution directly, rather than through the vehicle of Fréchet differentiability. Again the arguments are complex.

Because of the large sample and robustness qualities thus obtained, the author recommends use of the MM-estimators for estimation and inference. Recommendations for using the MM-estimator can be found in the package robustbase. For a review see Finger (2010).

To search out the outliers in large samples, the adaptive trimmed likelihood location estimator can be employed on the residuals obtained from the result of carrying out an MM-estimation, for example. Examples of this application are included in Bednarski et al. (2010) where the outliers identified are the same ones identified in the discussion of Section 7.4 for the wood-specific gravity data, the telephone data, and the Scottish hill races data. With this method the stack loss data identifies a different sequence of potential outliers and also identifies two outliers as observations 21 and 4. Bednarski et al. go on to describe the asymptotic normality of the resultant regression estimator obtained from the subsequent TLE under assumptions of heavy tailed symmetric errors. The result is given without proof in the next subsection.

7.5.1   TLE Asymptotics for Location and Regression

The idea that one chooses a proportion to minimize an estimate of a residual measure of variance, emanating originally from the study of Tukey and McLaughlin (1963) and solidified by Jaeckel (1971a), assumes that there is some reasonable convergence of the resulting estimator at the minimizing proportion to its asymptotic distribution. For the TLE it is reasonable to consider, at least for underlying smooth symmetric cumulative distributions F, because of empirical processes, that the minimizing proportion of $v\left(\alpha, F_n\right)$ in Eq. (7.19) is consistent to the minimizing proportion $v\left(\alpha, F\right)$, provided that proportion, call it α_A, occurs in the interior of a set $\left(\alpha_0, \alpha_1\right)$, where $\alpha_0 > 0$ and $G^*\left(n\right)/n \to \alpha_1$ and $\alpha_1 \leq \frac{1}{2}$. This is established under some conditions in Bednarski and Clarke (2002). It is conjectured in Bednarski and Clarke (2002) that it may also be the case that the minimizing value of $v\left(\alpha, F_n\right)$ converges to the minimizing value of $v(\alpha, F)$ which is zero when F_n is generated by $F = \Phi\{\cdot/\sigma\}$, but this is not shown in that paper. Bednarski and Clarke go on to show, using assumptions from Billingsley (1968) book on empirical processes, that the minimizing $\hat{\alpha}$ satisfies

$$\hat{\alpha} \to_p \alpha_A$$

and the location TLE $T_{\hat{\alpha}}[F_n]$ at that adaptive estimated proportion satisfies

$$\sqrt{n}\left(T_{\hat{\alpha}}[F_n] - \mu\right) \to_d N\left(0, V\left(\alpha_A, F\right)\right). \tag{7.27}$$

Whether one uses the minimizer of v or v^* does not make a difference, the result being only a change in α_A. Bednarski and Clarke illustrate the curve $v(\alpha, F)$ for the Laplace distribution. Similar illustrations can be made for other possible heavy tailed distributions exhibiting a unique minimum at a value of α_A.

As seen for regression, the computing time for an adaptive TLE would be prohibitively expensive for large n as $\binom{n}{g}$ soon explodes as g and n get large with g near $n/2$. However, choosing an initial estimator β^* to be a high breakdown point efficient estimator such as an MM-estimator one defines the residuals from MM-estimation as

$$r_i^* = Y_i - x_i^T \beta^*; i = 1, \ldots, n$$

and forms the ordered squared residuals

$$\left(r^*\right)^2_{1:n} \leq \left(r^*\right)^2_{2:n} \leq \cdots \leq \left(r^*\right)^2_{n:n}, \tag{7.28}$$

from which one calculates an estimate $V_n^*(g)$ using Eq. (7.24) so that

$$V_n^*(g) = \mathrm{var}\left(\tilde{\sigma}_*^2(g), \frac{g}{n}\right),$$

where

$$\tilde{\sigma}_*^2(g) = \frac{1}{n-p} \sum_{i=1}^{h}\left(r^*\right)^2_{i:n} \tag{7.29}$$

and $h = n - g$, while $\alpha = g/n$. Here $V_n^*(g)$, keeping the proportion $g/n = \alpha$ fixed, is a consistent estimate of the asymptotic variance of the innovation errors. The approach is to minimize $V_n^*(g)$ over $0 \leq g \leq G^*(n)$, where $G^*(n) = n - [\frac{n}{2}] - [\frac{p+1}{2}]$ to determine

$$V_n^*(\tilde{g}) = \min_{0 \leq g \leq G^*(n)} V_n^*(g).$$

The quantity $\tilde{S}_n$ is the set of $\tilde{h} = n - \tilde{g}$ indices in $\mathbf{r}^*$ that give the smallest $(r^*)_{i:n}^2$. The modified adaptive TLE of the vector parameter β is

$$\tilde{\beta} = \left(\tilde{X}^T\tilde{X}\right)^{-1}\tilde{X}^T\tilde{Y},$$

where $\left(\tilde{X},\tilde{Y}\right)$ are the original (X,Y) made up with indices in $\tilde{S}_n$, i.e. the least squares estimator with the "outliers" omitted.

Note that we no longer require the value of $\tilde{\sigma}^2(g)$ which is calculated using the above computationally intensive algorithm for brute force ATLA. It can be remarked that while the algorithm is not as computationally intensive as before, since it only uses $\tilde{\sigma}_*^2(g)$, the algorithm does require the evaluation of the MM-estimator. However, we can be confident in the light of the publicity given to MM-estimation that the estimation is feasible for reasonable sample sizes. The paper of Bednarski et al. (2010) discusses the asymptotics of the convergence when the errors have a symmetric density that is fatter tailed than normal (also called heavy-tailed), and then ideally we would find that for $\tilde{\alpha} = \tilde{g}/n$ we have

$$\tilde{\alpha} \to_p \alpha_A$$

and the regression modified TLE satisfies

$$\sqrt{n}\left(\tilde{\beta} - \beta\right) \to_d N\left(0, V\left(\alpha_A, F\right)\left(\mathbf{X}^T\mathbf{X}\right)^{-1}\right).$$

It can be remarked that one can use any suitable initial high breakdown estimator and not necessarily stick to MM-estimators. The tau estimators of Yohai and Zamar (1988) discussed in Maronna et al. (2006, section 5.14.1) are also possible. One can also use LTS and S-estimators as initial estimates.

The results alluded to in this section show at least the estimators are regular in a sense that the resulting estimators are asymptotically normal. However, the regularity conditions required by a noncontinuous J_α trimming function are stringent. Results are only for suitably smooth distribution functions for the errors. These are unrealistic in the true robustness framework set up by Hampel, and the MM-estimators would better provide the vehicle for robust inference. But knowing the potential "outliers" from MM-estimation is also worthy of consideration, whence the methodology of adaptive trimming of the residuals provides a useful avenue for discussion. When n becomes unusually large, search algorithms using random sampling are typically employed. See for example Salibian-Barrera et al. (2008) and Salibian-Barrera and Yohai (2012). One can never be sure one has all the outliers in any one run in such implementations, but with high probability the "outliers" can be found.

PROBLEMS

7.1. The estimator y_{T_g} assumes that one chooses to trim g observations from each tail prior to calculating the estimator. Thus, for example, a 5% trimmed mean when $n = 20$ trims g equal to one observation from each tail. On the other hand, the trimmed mean is not yet defined when g is not an integer. This can however be remedied. Let $\{y_i\}_{i=1}^n$ be the order statistics from a batch of n values, $0 < \alpha < 1/2$, and $p = 1 + [\alpha n] - \alpha n$, then the α-trimmed mean can be defined as

$$y_T = \frac{1}{n(1 - 2\alpha)} \{py_{[\alpha n]+1} + y_{[\alpha n]+2} + \cdots + y_{n-1-[\alpha n]} + py_{n-[\alpha n]}\}.$$

The α-trimmed mean attempts to discard the extreme $100\alpha\%$ observations from the averaging process. On the other hand, the α-Winsorized mean y_W attempts to give the same weight to the lower $100\alpha\%$ observations as the 100αth percentile, and the upper $100\alpha\%$ observations are given the same weight as the $100(1 - \alpha)$th percentile.

(a) Give a similar definition for the α-Winsorized mean, as for y_T.

(b) Write a program (e.g., in R or MATLAB) to evaluate, for arbitrary trimming proportion $0 < \alpha < 1/2$, the Winsorized mean of part (a).

(c) Use the program you have written in part (b) and the data set from Cushny and Peebles (1905) of section 7.3 to evaluate the 17% Winsorized mean, and the 10% Winsorized mean. Check your calculations using a scientific calculator.

7.2. Many L-estimators can be represented in a general way via

$$T[F_n] = \int_0^1 F_n^{-1}(t) J(t)\, dt.$$

Writing $G = (1 - \epsilon) F + \epsilon \delta_x$, where

$$\delta_x(y) = \begin{cases} 1 & \text{if } y \geq x \\ 0 & \text{Otherwise} \end{cases}$$

Now either considering problem 9 of section 2.1 or referring elsewhere such as Staudte and Sheather (1990, problem 10, p. 88) see that

$$\frac{\partial}{\partial \epsilon}\left(G^{-1}(t)\right)\big|_{\epsilon=0} = \frac{t - \delta_x\left(F^{-1}(t)\right)}{f\left(F^{-1}(t)\right)},$$

where $f(t) = F'(t)$.

(a) Since IF $(x, F, T) = \frac{d}{d\epsilon} T[G]|_{\epsilon=0}$ deduce that

$$
\text{IF}(x, F, T) = \int_0^1 \frac{tJ(t)}{f\left(F^{-1}(t)\right)} dt - \int_{F(x)}^1 \frac{1}{f\left(F^{-1}(t)\right)} J(t) dt
$$

assuming $F'(x) = f(x)$ exists and is continuous.

(b) Show that

$$
\text{IF}(x, F, T) = -\int_{-\infty}^{+\infty} (I(y \geq x) - F(y)) J(F(y)) dy,
$$

where $I(A)$ is the indicator function of the set A. This shows that

$$
\text{IF}(x, F, T) = \int_{-\infty}^{+\infty} F(u) J(F(u)) du - \int_x^\infty J(F(u)) du.
$$

i. Suppose

$$
T[F] = \sum_{i=1}^m \beta_i F^{-1}(t_i) \qquad 0 < t_1 < \cdots < t_m \leq 1
$$

and for appropriate choice of $\beta_i's$ $T[F] = \mu$. Then

$$
T[F_n] \approx \sum_{i=1}^m \beta_i X_{(t_i n)}.
$$

This is approximate since $t_i n$ is not always an integer. Write down the influence function of this estimator.

ii. In the immediate previous part of this question substitute $J(u) = 1$, $t_1 = \frac{1}{2}$, and $\beta_1 = 1$. What is your estimator and what is its influence function?

iii. In the case where $J(t) = \frac{1}{1-2\alpha} I([\alpha, 1-\alpha])$ describe your estimator and derive the influence function. This estimator is discussed in many texts and it may be worthwhile comparing other sources.

8

TRIMMED LIKELIHOOD FOR MULTIVARIATE DATA

8.1 IDENTIFICATION OF MULTIVARIATE OUTLIERS

The minimum covariance determinant (MCD) estimator lies at the heart of trimmed likelihood methodology since it is the harbor of least trimmed squares in the univariate location model, and the least trimmed squares estimator as has been illustrated is just one example of the trimmed likelihood estimator. There is now a large literature on the MCD estimator as can be gleaned from Hubert and Debruyne (2009). Rousseeuw (1983, 1984) defined the location MCD-estimator more generally as a multivariate estimator, the asymptotic theory for which was detailed in Butler et al. (1993). It was using this description of asymptotic theory and the knowledge of adaptive trimmed likelihood methodology in location and regression that led Clarke and Schubert (2006) to describe variants of an adaptive trimmed likelihood algorithm to identify outliers in a multivariate sample. Much but not all of the discussion in this section is with permission from that paper. In the univariate setting, most outliers can be exposed by elementary box-whisker plots or stem and leaf plots and in two dimensions scatter plots usually are sufficient. On the other hand in multivariate estimation, visual inspections are not possible in greater than three dimensions and even three-dimensional data sets can contain outliers that evade detection from certain angles and so one is dependent on analytical methods for outlier detection.

Robustness Theory and Application, First Edition. Brenton R. Clarke.
© 2018 John Wiley & Sons, Inc. Published 2018 by John Wiley & Sons, Inc.
Companion website: www.wiley.com/go/clarke/robustnesstheoryandapplication

Here assuming a multivariate normal distribution, the equivalent of the adaptive trimmed likelihood estimator (ATLA) is to choose that multivariate MCD estimate for location with a trimming proportion that minimizes the estimate of the determinant of the asymptotic variance of the MCD estimate. The aim is to trim less than approximately 50%.

Clarke and Schubert's proposal involves starting with a robust MCD estimate for centroid and scale followed by what is referred to as a Forward Search Algorithm, first developed by Hadi (1992, 1994), Atkinson (1994), Rocke and Woodruff (1996), which, in the spirit of Atkinson and Riani (2000) and Atkinson et al. (2004), searches for a subset of data, at each stage of the Forward Search, resulting in the minimum determinant of the estimated asymptotic variance. The theoretical premise for this proposal, as described in the previous motivations for ATLA, is based on the principle that trimming a normally distributed data set will necessarily reduce the information it contains. This reduction in information increases the estimated variability of the location estimate unless those observations trimmed are outliers.

Consider any set of independent identically distributed observations $\{X_i\}_{i=1}^n$ in p-dimensional Euclidean space $\mathcal{E}^p$ and let S_n be an arbitrary subset with the number of observations in that subset equal to $h_n = \lfloor n\gamma \rfloor$ for $\gamma = 1 - \alpha$, where α corresponds to the amount of trimming, $0.5 < \gamma \leq 1$. As in the literature cited, we denote $T[F_n]$ to be the univariate estimate of location, and then correspondingly we can denote $T[F_n]$ to be the multivariate estimate of location. Define in the spirit of Butler et al. (1993),

$$\hat{\mathcal{M}}\{S_n\} = \frac{1}{h_n} \sum_{i \in S_n} X_i,$$

and

$$\hat{S}\{S_n\} = \frac{1}{h_n} \sum_{i \in S_n} (X_i - \hat{\mathcal{M}}\{S_n\})(X_i - \hat{\mathcal{M}}\{S_n\})^T.$$

Now consider the subset $\hat{S}_n$ for which the determinant of $\hat{S}\{S_n\}$, $|\hat{S}\{S_n\}|$, attains its minimum value over all subsets of S_n having h_n observations. The estimator $\hat{\mathcal{M}}\{\hat{S}_n\}$ is the MCD estimator of Rousseeuw (1983), which was defined in terms of the perhaps more easily understood notation writing $T[X] \equiv T[X_1, \dots, X_n]$ and realizing $h \equiv h_n$ is used in the previous chapter.

$T[X] \; : = $ mean of the h points of X for which the determinant

of the covariance matrix is minimal.

The precise form for the location and scatter functional and their influence functions is described in Cator and Lopuhaä (2012). Clarke and Schubert (2006) presumed rightly the multivariate location estimator could be written as a functional

$T[\cdot]$ evaluated at the empirical distribution function, F_n, which assigns atomic mass $\frac{1}{n}$ to each of the points X_i in the sample. A bold T is referred to in Clarke and Schubert to denote the vector location functional which we use now.

When analyzing multivariate data sets we are dealing with elliptical probability densities of the form

$$\frac{1}{|\Sigma|^{1/2}} h^*((\mathbf{x} - \boldsymbol{\mu})^T \sum{}^{-1}(\mathbf{x} - \boldsymbol{\mu})), \tag{8.1}$$

where $h^* : \mathcal{E}^+ \to \mathcal{E}^+$ is assumed to be nonincreasing, yielding a unimodal density (Butler et al., 1993), with $\boldsymbol{\mu}$ the centroid, $\sum$ the covariance matrix, and $|\sum|$ denoting the determinant of $\sum$ which is assumed to be nonzero.

If the sample data are from a multivariate normal distribution with mean zero and covariance matrix the identity, $N(\mathbf{0}, \mathbf{I}_p)$, then the sample covariance matrix using the MCD estimator of Butler et al. (1993) is such that

$$\hat{\Sigma}_\alpha[F_n] \xrightarrow{wp1} \rho(\gamma)\mathbf{I}_p.$$

The form for $\rho(\gamma)$ is discussed in Clarke and Schubert (2006).

With regard to our sample estimate of the multivariate mean, $T[F_n]$, we have

$$\sqrt{n}(T[F_n] - \boldsymbol{\mu}) \xrightarrow{d} N(0, \kappa(\gamma)\mathbf{I}_p),$$

again the form of $\kappa(\gamma)$ being discussed in Clarke and Schubert.

If our data were exactly from a multivariate normal distribution, the $\kappa(\gamma)$ above could be used to give an estimate for the asymptotic variance of $T[F_n]$ assuming $\gamma = 1 - \alpha$, where α is the proportion of trimming. If our sample data consist of outliers, then one would expect the value of $\rho(\gamma)\mathbf{I}_p$ to disagree with our sample variance $\hat{\Sigma}_\alpha[F_n]$, which is no longer a covariance matrix from a normal distribution. With this in mind the multivariate extension of minimizing Eq. (7.19) becomes, as in Clarke and Schubert, choosing γ (equivalently choosing α) to minimize

$$\left|\frac{\kappa(\gamma)}{\rho(\gamma)}\hat{\Sigma}_\alpha[F_n]\right| = \frac{|\hat{\Sigma}_\alpha[F_n]|}{\left(\frac{4\pi p/2}{p\Gamma(p/2)} \int_0^{r_\gamma} r^{p+1} q'(r^2) dr\right)^{2p}}. \tag{8.2}$$

where $q(u) = (1/(2\pi))^{p/2} e^{-u/2}$ and $r_\gamma^2 = \chi^2_{p;1-\alpha}$. Here $\chi^2_{p;1-\alpha}$ satisfies $1 - \alpha = p(W < \chi^2_{p;1-\alpha} | W \sim \chi^2_p)$. Γ is the usual gamma function, $\Gamma(v) = \int_0^\infty s^{v-1} e^{-s} ds$. In fact minimizing formula (8.2), which we will call Type 1, or T1, is equivalent to minimizing $v(\alpha, F_n)$ when $p = 1$, which is the preferred option 4 in Clarke

(1994). By Bednarski and Clarke (1993) for univariate data, $p = 1$ and $F = \Phi$, the distribution function of the standard normal variable, we see that

$$(1 - \alpha)\hat{\Sigma}_\alpha[F_n] \xrightarrow{wpl} \int_{-z_{\alpha/2}}^{z_{\alpha/2}} x^2 \, d\Phi(x).$$

This yields the Fisher consistent estimate for Σ which in the multivariate setting would result in choosing γ to minimize our Type 2 or T2 proposal

$$\left| \frac{\kappa(\gamma)}{\rho(\gamma)} (1 - \alpha)\hat{\Sigma}_\alpha[F_n] \right|. \tag{8.3}$$

For $p = 1$, minimizing the objective function in Eq. (8.3) is equivalent to choosing option 5 in Clarke (1994).

The implementation of proposals T1 and T2 is described in Clarke and Schubert (2006) and more multivariate applications can be gleaned from Schubert (2005). An algorithm in R developed by the author and Daniel Schubert and which has been slightly modified with acknowledgment for ease of use is made available. See Problem 8.1.

When generating clean data sets generated from multivariate normal distributions, the authors discovered in simulations not all reported that proposal T2 was too sensitive for sample sizes $n \leq 30$ but performed better than T1 for larger samples. With this insight, we decided to apply the proposal T1 to samples of $n = 20$ and to use T2 for the other Monte Carlo samples with $n \geq 30$. Daniel Schubert's favorite data set used as vindication of the method concerned a multivariate data set of four variables recorded by a set of English and Australian batsmen in what some may say is that "quaint" game of cricket. This game, however, captures the imagination and essence of being for the average person in the street when played between the two countries of England and Australia. The four variables measured were number of innings played, x_1, number of hundreds, x_2, and fifties, x_3, scored and the number of runs amassed, x_4, by the top 90 batsmen. The analysis extracted only one batsman out as an outlier and as every cricket follower would know, at least in these two countries, that was "Don Bradman."

Naturally, a query would be about statistics of baseball. For instance, is Babe Ruth standing out as the best? Such an assessment may be harder to make because of all the seemingly great baseballers. Choosing the right variables to measure and the time period would be an interesting study.

The method T1 was applied to the multivariate data set of Table 7.1, treating the five independent variables, and the dependent variable, as a multivariate set of dimension six. Despite this being a different scenario to the regression situation, it occurred that the same observations, 4, 6, 8 and 19 showed up as outliers.

It should be pointed out that there are now suggestions that allow deterministic evaluation of the MCD estimator. For example, compare with Hubert et al.

(2012). These potentially can be allied to the adaptive trimmed likelihood algorithm search.

There are more recent expositions that use similar forward search methods as is used in Clarke and Schubert (2006). These include, as mentioned above, Atkinson et al (2004), and a more recent innovation involving monitoring and adaptive estimation by Cerioli et al. (2017). Another classical discussion is given in Hubert et al. (2008).

PROBLEMS

8.1. Using the program made available at the ancillary materials website for this book, apply the ATLA multivariate algorithm to the Wood Volume Data and confirm the outliers recorded in Table 7.1.

8.2. Consider the Cricket data at that site. Run the algorithm and confirm "Bradman" as the only outlier.

9

FURTHER DIRECTIONS AND CONCLUSION

9.1 A WAY FORWARD

Seemingly, we began with a methodology that pointed us in the direction of M-estimators with bounded continuous influence functions for in them is a natural continuity of one's estimate in the face of possible small aberrations in either the data or departures from the model distribution in the underlying distribution that generates the data. However, these last chapters emphasize a different type of estimation method, the trimmed likelihood estimator (TLE), an adaptive version of which can be used to pinpoint outlying observations in location estimation, regression estimation, and multivariate estimation. The adaptive version is called the adaptive trimmed likelihood estimator or ATLA for short. It is strongly emphasized that the trimmed likelihood estimator for the normal distribution has an influence function that is the same as the Huber skipped mean. It is not continuous and therefore not Fréchet differentiable. Contamination near the points of discontinuity can lead to unstable behavior. If the main objective is to obtain a robust estimate with stable inferences allowing for the possibility of a small proportion of contamination in one's data, or slight departures from the model parametric family, then using a Fréchet differentiable estimator with associated robust asymptotic theory for confidence intervals and tests is sensible.

Robustness Theory and Application, First Edition. Brenton R. Clarke.

In regression, for example, a natural estimator to use is the MM-estimator for which there is supporting asymptotic theory in the literature. Applying adaptive trimmed likelihood algorithms to the residuals from MM-estimation can be a natural approach to identifying any outlying observations. For larger proportions of contamination in the range of 30–50% things are not so clear. Applying the MM-estimator with default parameters can lead in some instances to an estimator that by eye clearly fails. The illustration from Figure 9.1 is on some data supplied from Campbell (2016). Here ostensibly there is 50% contamination. There are two clearly defined groups of data, which overlap. Applying the adaptive trimmed likelihood algorithm of Clarke and Schubert (2006) to the bivariate data leads to the maximum proportion of identifying 4999 outliers out of 10 000 points. The least squares regression line fit on the remaining data is illustrated demonstrating that at least one is gaining a reasonable fit to the population of interest. On the other hand, it appears that MM-estimator in regression using the default parameters in version 3.3.1 of R appears to fail in that it straddles the two subpopulations, that with the upper population of "good" data and that with the lower subpopulation of "bad" data.

Clearly, there is a need to be aware of such situations especially if there is the likelihood of larger proportions of contamination, and this is more important with multivariate data and "Big Data" where it is not easy to visualize the contamination. This is not to say we should do away with MM-estimators, rather, we should recognize that it offers continuous inferences in small neighborhoods of the model, which is certainly an advantage over classical estimation methods.

Are there models where the trimmed likelihood estimator is weakly continuous and Fréchet differentiable? The answer to this is explored in Clarke et al. (2000, 2011). There it is shown for the negative exponential distribution that the resulting trimmed likelihood estimator for a positive β proportion of trimming is the β-trimmed mean. This estimator is discussed broadly in the book by Staudte and Sheather (1990). The estimator is weakly continuous at the model distribution and, as shown in Clarke et al. (2000), is also Fréchet differentiable at least at the parametric model.

More complex modeling can be carried out using the trimmed likelihood estimator, however, while the estimation methods involve trimming and using the likelihood equations, real questions are to be asked about the Fisher consistency, since one is trimming a proportion of the data before estimation. One philosophy is to apply the trimmed likelihood estimation method when there is known to be contamination in the data. For example, in the study of mixtures see Neykov et al. (2007) and Gallegos and Ritter (2009). The question remains, when the model is a correctly specified mixture of normal distributions say, does one gain consistent estimates? This warrants more discussion.

For a recent discussion of related lifetime models with exponential or log normal distributions possessing a linear or nonlinear link functions, the implicit function theorem is used to derive the influence function of the trimmed likelihood

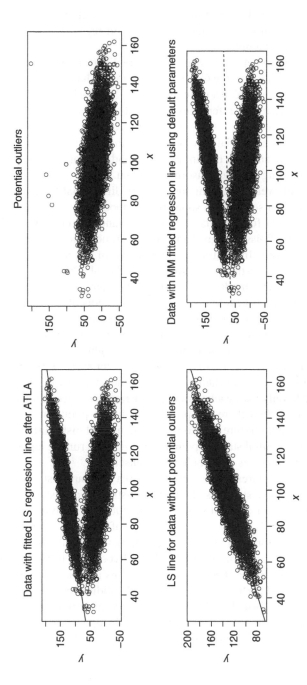

FIGURE 9.1 These plots are formed from 10 000 bivariate data supplied by Norm Campbell, 5000 points from each of two populations. A plot of bivariate data with the least squares fitted line on the data after the multivariate ATLA algorithm of Clarke and Schubert (2006) is applied. A plot of the potential outliers identified by the multivariate ATLA and the plot of the least squares line with the observations retained by ATLA. The final plot shows that the MM-algorithm fitted regression dashed line applied with default parameters in version 3.2.2 of R. (*Source:* Clarke and Schubert (2006). Reproduced with the permission of John Wiley & Sons.)

estimators in quite complex models in Müller et al. (2016). In addition, these corroborate the earlier derivation of the influence function in regression seen from Bednarski et al. (2010), for example.

There are other problems where robustness plays a real part. In the area of survival analysis, the robustification of the Cox regression model has been examined in Bednarski (1993b) assuming Fréchet differentiability. See the book by Heritier et al. (2009) for some more recent references. A recent study of lifetime models with censored data by Clarke et al., (2017) shows gains can be made in terms of smaller mean squared errors in estimation of mean lifetimes by choosing a hybrid estimator of a β-trimmed mean and a maximum likelihood estimator that depends on the proportion of trimming and the censoring constant.

Recent applications of weakly continuous Fréchet differentiable estimating functionals have appeared in some quite sophisticated models. For example, see Kulawik and Zontek (2015) for models including variance components. Also there are models involving Poisson regression and also time series in Bednarski (2004) and (2010) respectively. Interesting theory in terms of risk analysis, which borrows from the theory of robustness, is found in Kiesel et al. (2016).

The more complicated the model the more likely there are going to be requirements for efficient algorithms which search for the roots of the estimating equations and/or the minima of the objective function. More frequently, it will be the case that searching for the consistent root of an equation or a global minimum of an appropriate objective function will need some insight into the physical problem at hand in order to propose regions of the parameter space to begin with an initial estimate. This can be important, for if the dimension of the parameter space is r and we wish to only vary each component parameter twice in the initial grid of search parameters, then there are possibly 2^r initial points in the grid. For example, if $r = 5$ as in a mixture of two univariate normal distributions then there would be a grid of 32 initial starting points from which to run the algorithm. Having more parameters increases the problem immensely.

In this book we have not addressed the very wide area of robust testing. There are serious questions about test statistics and their robustness. For instance, what is the effect on level and power of a test statistic by an outlier or outlying values? Hampel et al. (1986, chapter 3, p. 187) cite the first use of the word "robustness" as by Box (1953) in connection with tests. For the uninitiated, the text by Staudte and Sheather (1990) addresses some of the pitfalls in hypothesis testing caused by outlying values in one's sample. The classic discussions include robustness of the t-statistic, which as noted earlier has "robustness of validity" but not "robustness of efficiency." A more formal basis to the influence of outliers on tests is offered in Hampel et al. (1986, section 3.1) who adopt the approach of an influence function for tests as defined by Rousseeuw and Ronchetti (1979, 1981), and it is interesting that in further development Heritier and Ronchetti (1994) illustrate their framework in terms of Fréchet differentiable functionals. The natural implications of robust estimates and standard errors of estimates, and consequent robust confidence intervals based on asymptotic distributions is afforded by Fréchet differentiable functionals. This then is a natural entry into robust testing.

Another side issue with respect to the current emphasis on confidence intervals rather than the use of p-values is raised in the current awareness of misuse of p-values in publication as raised, for example, in Baker (2016). She comments particularly on the article of Wasserstein and Lazar (2016) on "The American Statistical Association's Statement on p-Values: Context, Process, and Purpose." However, in response to ideas of robustness Wilcox and Serang (2017) suggests that we do not throw out p-values all together.

PROBLEM

9.1. Consider the β-trimmed mean which is a solution of equations that mimic equation (7.12) except this time the parametric distribution is the exponential distribution with $F_\theta(x) = 1 - \exp(-x/\theta)$, $x > 0$, $\theta > 0$. For example, the equations are

$$L(G, \theta) = \int \phi(x, \theta) J_\beta[G\{y : \log f_\theta(y) \leq \log f_\theta(x)\}] dG(x) = 0, \quad (9.1)$$

where $\phi(x, \theta) = (\partial/\partial\theta) \log f_\theta(x)$. Show that so long as the distribution G puts zero weight at the point $G^{-1}(1 - \beta)$, then the estimator functional evaluated at the distribution G can be written as

$$T_\beta[G] = \frac{1}{1 - \beta} \int_0^{G^{-1}(1-\beta)} x dG(x) = \frac{1}{1 - \beta} \int_0^{1-\beta} G^{-1}(t) dt.$$

Show that this estimator has a bounded continuous influence function at the model parametric family distribution.

APPENDIX A

SPECIFIC PROOF OF THEOREM 2.1

Lemma A.1: *Lemma 6.1 in Clarke (1983a) Let $A \in \mathcal{B}$ be a continuity set of F_θ and $\kappa > 0$ be given. Then there exists an $\epsilon > 0$ such that for any $G \in \mathcal{G}$,*

$$d_p(G, F_\theta) < \epsilon \quad \text{implies} \quad |G\{A\} - F_\theta\{A\}| < \kappa.$$

Proof. Since A is a continuity set, there is some $\epsilon_1 > 0$ for which $F_\theta\{A^{\epsilon_1}\} < F_\theta\{A\} + \kappa/2$. Choose $\epsilon_1 < \kappa/2$. Then if $d_p(G, F_\theta) < \epsilon_1$,

$$G\{A\} < F_\theta\{A^{\epsilon_1}\} + \epsilon_1 < F_\theta\{A\} + \kappa.$$

Similarly, since $R - A$ shares the same boundary of A, there is some $\epsilon_2 > 0$ for which

$$F_\theta\{A^{-\epsilon_2}\} = F_\theta\{R - (R - A)^{\epsilon_2}\} > F_\theta\{A\} - \frac{\kappa}{2}.$$

Choose $\epsilon_2 < \kappa/2$ and suppose $d_p(G, F_\theta) < \epsilon_2$. Then

$$G\{A\} > F_\theta\{A^{-\epsilon_2}\} - \epsilon_2 > F_\theta\{A\} - \kappa.$$

The lemma follows by setting $\epsilon = \min(\epsilon_1, \epsilon_2)$ $\square$.

Robustness Theory and Application, First Edition. Brenton R. Clarke.
© 2018 John Wiley & Sons, Inc. Published 2018 by John Wiley & Sons, Inc.
Companion website: www.wiley.com/go/clarke/robustnesstheoryandapplication

Proof. (**of Theorem 2.1 which is Theorem 6.1 in Clarke (1983a)**)

Since R is separable and complete, there exists a compact set C such that $F_\theta\{R - C\} < \delta/(16.H)$, which further can be chosen to be a continuity set of F_θ. For arbitrary $\eta > 0$, by Lemma 3.1 of Rao (1962), there exists a finite number of sets $\{A_j\}_{j=1}^n$, where $n = n(\eta)$, such that (a) $\cup_{j=1}^n A_j = C$; (b) $A_j \cap A_{j'} = \emptyset$ for $j \neq j'$; (c) for each j, A_j is a continuity set for F_θ; and (d) for any $x, y \in A_j$ and $f \in \mathcal{A}$, $|f(x) - f(y)| < \eta$, for each $j = 1, \ldots, n$. Let $\eta = \delta/4$ and choose $\{y_j\}_{j=1}^n$ arbitrarily in $\{A_j\}_{j=1}^n$ respectively and let F_θ^* be the possibly improper measure attributing weight $F_\theta\{A_j\}$ to the point y_j, for each, $j = 1, \ldots, n$. Then for each $f \in \mathcal{A}$

$$|\int_C f dF_\theta - \int_C f dF_\theta^*| \leq \sum_{j=1}^n \int_{A_j} |f(x) - f(y_j)| dF_\theta < \frac{\delta}{4}.$$

Hence,

$$\sup_{f \in \mathcal{A}} |\int_C f dF_\theta - \int_C f dF_\theta^*| \leq \frac{\delta}{4}.$$

Similarly, given $G \in \mathcal{G}$ let G^* be that measure attributing weight $G\{A_j\}$ to y_j for each $j = 1, \ldots, n$. Then

$$\sup_{f \in \mathcal{A}} |\int_C f dG - \int_C f dG^*| < \frac{\delta}{4}.$$

Now,

$$|\int_C f dF_\theta^* - \int_C f dG^*| \leq H \sum_{j=1}^n |F_\theta\{A_j\} - G\{A_j\}|.$$

By Lemma A.1 choose ϵ_j such that $G \in \mathcal{G}$, $d_p(G, F_\theta) < \delta_j$ implies $|F_\theta\{A_j\} - G\{A_j\}| < \delta/(4.n.H)$. Let ϵ_0 be so that $G \in \mathcal{G}$, $d_p(G, F_\theta) < \epsilon_0$ implies $|F_\theta\{R - C\} - G\{R - C\}| < \delta/(16.H)$. Set $\epsilon = \min(\epsilon_0, \epsilon_1, \ldots, \epsilon_n)$. Then $G \in \mathcal{G}$, $d_p(G, F_\theta) < \epsilon$ implies

$$\sup_{f \in \mathcal{A}} |\int f dF_\theta - \int_R f dG| \leq \sup_{f \in \mathcal{A}} |\int_{R-C} f dF_\theta - \int_{R-C} f dG|$$

$$+ \sup_{f \in \mathcal{A}} |\int_C f dF_\theta - \int_C f dF_\theta^*| + \sup_{f \in \mathcal{A}} |\int_C f dG - \int_C f dG^*|$$

$$+ \sup_{f \in \mathcal{A}} |\int_C f dF_\theta^* - \int_C f dG^*|$$

$$< H.[F_\theta\{R - C\} + G\{R - C\}] + \frac{1}{4}\delta + \frac{1}{4}\delta + \frac{1}{4}\delta < \delta.$$

$\square$

APPENDIX B

SPECIFIC CALCULATIONS IN EXAMPLES 4.1 AND 4.2

Here we give some of the results supporting the calculations made in Section 4.3.

Consider firstly the Cauchy location parametric model for which $f_\tau(x) = \frac{1}{\pi(1+(x-\tau)^2)}$, whence the efficient score function is $\psi(x, \tau) = \frac{\partial}{\partial \tau} \ln f_\tau(x) = \frac{2(x-\tau)}{1+(x-\tau)^2}$ and $\frac{\partial}{\partial \tau} \psi(x, \tau) = \frac{-2+2(x-\tau)^2}{[1+(x-\tau)^2]^2}$. Then GJ's selection functional defined by

$$|\hat{SD}_1^{-2} - \hat{SD}_2^{-2}| = |\int \psi(x, \tau)^2 dF_n(x) + \int \frac{\partial}{\partial \tau} \psi(x, \tau) dF_n(x)|$$

$$= |\int \frac{6(x-\tau)^2 - 2}{[1 + (x-\tau)^2]^2} dF_n(x)|.$$

Similarly, one can establish formula (4.24). Now consider our Example 4.2. Here $f_\tau(x) = \frac{1}{\sqrt{2\pi}\tau} \exp(\frac{-(x-\tau)^2}{2\tau^2})$. It then follows after some calculation (see Appendix C) that the first eight moments of the distribution $f_\theta(x)$ where $\mu_k = E_\theta[X^k]$ are $\mu_1 = \theta$, $\mu_2 = 2\theta^2$, $\mu_3 = 4\theta^3$, $\mu_4 = 10\theta^4$, $\mu_5 = 26\theta^5$, $\mu_6 = 76\theta^6$, $\mu_7 = 232\theta^7$, and $\mu_8 = 764\theta^8$. Now let us examine $\tilde{\psi}_{GJ}(x, \tau) = \frac{2}{\tau^2} + \frac{4x}{\tau^3} - \frac{4x^2}{\tau^4} - \frac{2x^3}{\tau^5} + \frac{x^4}{\tau^6}$. It can be noted $E_\tau[\tilde{\psi}(X, \tau)] = 0$. To evaluate $J_\theta^{GJ}(\tau)$, we see $\frac{\partial}{\partial \tau} \tilde{\psi}_{GJ}(x, \tau) = \frac{-4}{\tau^3} + \frac{-12x}{\tau^4} + \frac{16x^2}{\tau^5} + \frac{10x^3}{\tau^6} + \frac{-6x^4}{\tau^7}$. Now $J_\theta^{GJ}(\tau) =$

Robustness Theory and Application, First Edition. Brenton R. Clarke.
© 2018 John Wiley & Sons, Inc. Published 2018 by John Wiley & Sons, Inc.
Companion website: www.wiley.com/go/clarke/robustnesstheoryandapplication

$E_\theta[\frac{\partial}{\partial\tau}\tilde{\psi}_{GJ}(X,\tau)] = \frac{-4}{\tau^3} + \frac{-12\theta}{\tau^4} + \frac{32\theta^2}{\tau^5} + \frac{40\theta^3}{\tau^6} + \frac{-60\theta^4}{\tau^7}$. Note then that $J_\theta^{GJ}(\theta) = \frac{-4}{\theta^3}$
while at the erroneous root $J_\theta^{GJ}(\theta_1) = J_\theta^{GJ}(-2\theta) = \frac{-5}{32\theta^3}$. Given ψ is the efficient
score function and $\psi(x,\tau) = \frac{-1}{\tau} + \frac{-x}{\tau^2} + \frac{x^2}{\tau^3}$ so that $\frac{\partial}{\partial\tau}\psi(x,\tau) = \frac{1}{\tau^2} + \frac{2x}{\tau^3} + \frac{-3x^2}{\tau^4}$
and $M_\theta(\tau) = \frac{1}{\tau^2} + \frac{2\theta}{\tau^3} + \frac{-6\theta^2}{\tau^4}$. It follows that $M_\theta(\theta) = \frac{-3}{\theta^2}$, while at the erroneous
root one obtains $M_\theta(\theta_1) = M_\theta(-2\theta) = \frac{-3}{8\theta^2}$. Now the formula for $\psi_2(x,\tau)$ given
by Eq. (4.29) which inherently depends on the choice of $\tilde{\psi}$ is for the choice
of GJ given as $\psi_{2GJ}(x,\tau) = \frac{5}{2\tau} + \frac{4x}{\tau^2} + \frac{-4x^2}{\tau^3} + \frac{-3}{2}\frac{x^3}{\tau^4} + \frac{3}{4}\frac{x^4}{\tau^5}$. Hence, $\psi_{2GJ}^2(x,\tau) = $
$\frac{31}{4}\frac{x^4}{\tau^6} - \frac{79}{2}\frac{x^3}{\tau^5} - 4\frac{x^2}{\tau^4} + 18\frac{x^5}{\tau^7} - \frac{15}{4}\frac{x^6}{\tau^8} - \frac{9}{4}\frac{x^7}{\tau^9} + \frac{9}{16}\frac{x^8}{\tau^{10}} + 20\frac{x}{\tau^3} + \frac{25}{4}\frac{1}{\tau^2}$.Consequently,
$E_\tau[\psi_{2GJ}^2(X,\tau)] = \frac{57}{2\tau^2}$. It can be noted $E_\tau[\psi_2(X,\tau)] = 0$. Since we are acting as
though the "true" τ is -2, we should get $\sigma_{\psi_{2GJ}}^2(-2,-2) = \frac{57}{8} = 7.125$ and the
standard error of our test statistic T_n^{GJ} would be $|\{\frac{-4}{-3}\}|\sqrt{57/2}/4 = 1.7795$. In
fact, the statistic

$$\frac{1}{\sqrt{n}}\sum_{j=1}^n \tilde{\psi}_{GJ}(X_j,\hat{\theta}_{1n}) \to_d N\left(\frac{-3\sqrt{n}}{32}, \left\{\frac{J_1(-2)}{M_1(-2)}\right\}^2 \sigma_{\psi_2}^2(1,-2)\right).$$

Note we substituted $\theta = 1$ here since the generating distribution is F_1. To evaluate
$\sigma_{\psi_{2GJ}}^2(1,-2)$, we note first that $E_1[\psi_{2GJ}(X,\tau)] = \frac{-9}{40}$. For $\theta = 1$ and $\tau = -2$ we
obtain

$$\psi_{2GJ}^2(x,-2) = \begin{array}{l} \frac{19}{80}x^2 + \frac{89}{80}x^3 - \frac{11}{1600}x^4 - \frac{171}{800}x^5 - \frac{21}{1600}x^6 + \frac{9}{800}x^7 + \\[2mm] \frac{9}{6400}x^8 - \frac{133}{100}x + \frac{49}{100} \end{array}$$

from which we find $E_1[\psi_{2GJ}^2(X,-2)] = \frac{1833}{1600}$. Then $\mathrm{var}_1[\psi_{2GJ}(X,-2)] = \frac{1833}{1600} - \frac{81}{1600} = \frac{1752}{1600}$. Now $\mu_\theta(\theta_1) = \frac{-3}{32\theta^2}$ and

$$\left\{\frac{J_1(-2)}{M_1(-2)}\right\}^2 \sigma_{\psi_{2GJ}}^2(1,-2) = \left\{\frac{-5/32}{-3/8}\right\}^2 \times \frac{1752}{1600} = \frac{73}{384}.$$

This verifies Eq. (4.36).

In a similar vein, we consider the test statistic

$$T_n^{CI} = \frac{1}{\sqrt{n}}\sum_{j=1}^n \tilde{\psi}_{CI}(X_j,\hat{\theta}),$$

where for the sake of preciseness we let

$$\tilde{\psi}_{CI} = -\frac{\partial}{\partial\tau}\psi(x,\tau) + \int\left\{\frac{\partial}{\partial\tau}\psi x,\tau)\right\} dF_\tau(x).$$

and again ψ is the efficient score function. This implies $\tilde{\psi}_{Cl}(x, \tau) = \frac{-4}{\tau^2} + \frac{3x^2}{\tau^4} - \frac{2x}{\tau^3}$.
Note $E_\theta[\tilde{\psi}_{Cl}(X, \theta)] = 0$. Also $\frac{\partial}{\partial \tau}\tilde{\psi}_{Cl} = \frac{8}{\tau^3} + \frac{-12x^2}{\tau^5} + \frac{6x}{\tau^4}$ from which we can derive
$J_\theta^{Cl}(\theta) = \frac{-10}{\theta^3}$ and $J_\theta^{Cl}(\theta_1) = J_\theta^{Cl}(-2\theta) = \frac{1}{8\theta^3}$. Now

$$\psi_{2Cl}(x, \theta) = \frac{-1}{5\theta} - \frac{x^2}{10\theta^3} + \frac{2x}{5\theta^2}$$

and

$$\psi_{2Cl}^2(x, \theta) = \frac{x^2}{5\theta^4} - \frac{2x^3}{25\theta^5} + \frac{x^4}{100\theta^6} - \frac{4x}{25\theta^3} + \frac{1}{25\theta^2}.$$

Then since $E_\theta[\psi_{2Cl}(X, \theta)] = 0$, it is not hard to see $\sigma_{\psi_{2Cl}}^2 = E_\theta[\psi_{2Cl}^2(X, \theta)] = \frac{3}{50\theta^2}$.
Believing that our erroneous consistent root $\hat{\theta}_{1n}$ to -2θ is in fact the true value,
we would on average have as our standard error for the test statistic

$$|J_{-2}(-2)/M_{-2}(-2)|\sigma_{\psi_{2Cl}}(-2, -2) = |\frac{10/8}{-3/4}|\sqrt{\frac{3}{50} \times \frac{1}{4}} = 0.2041 \ .$$

But in fact

$$\frac{1}{\sqrt{n}}\sum_{j=1}^n \tilde{\psi}_{Cl}(X_j, \hat{\theta}_{1n}) \to_d N\left(-\frac{3\sqrt{n}}{8}, \left\{\frac{J_1(-2)}{M_1(-2)}\right\}^2 \sigma_{\psi_{2Cl}}^2(1, -2)\right).$$

We can note $\mu_{Cl}(\theta, -2\theta) = E_\theta[\tilde{\psi}_{Cl}(X_j, -2\theta)] = \frac{-3}{8\theta^2}$. Then also

$$E_\theta[\psi_{2Cl}(X, -2\theta)] = -E_\theta[\psi(X, -2\theta)] + \left\{\frac{M_\theta(-2\theta)}{J_\theta(-2\theta)}\right\} E_\theta[\tilde{\psi}(X, -2\theta)]$$

$$= 0 + \left[\frac{\frac{-3}{8\theta^2}}{\frac{1}{8\theta^3}}\right] \times \frac{-3}{8\theta^2} = \frac{9}{8\theta}.$$

Now when $\theta = 1$

$$\psi_{2Cl}(x, -2) = \frac{5}{2} - \frac{x}{2} - \frac{7x^2}{16}$$

whereupon

$$\psi_{2Cl}^2(x, -2) = \frac{7x^3}{16} - \frac{31x^2}{16} + \frac{49x^4}{256} - \frac{5x}{2} + \frac{25}{4},$$

and so $E_\theta[\psi_{2Cl}^2(X, -2)] = 453/128$ and $\text{var}_1[\psi_{2Cl}(X, -2)] = \frac{453}{128} - \left\{\frac{9}{8}\right\}^2 = 291/128$. Now in fact

$$\frac{1}{\sqrt{n}}\sum_{j=1}^n \tilde{\psi}_{Cl}(X_j, \hat{\theta}_{1n}) \to_d N\left(\frac{-3\sqrt{n}}{8}, 0.5026^2\right),$$

which verifies Eq. (4.35), since $0.5026 = \sqrt{\frac{97}{384}}$ and one can verify the standard error of the test statistic is

$$\left|\frac{J_1(-2)}{M_1(-2)}\right| \sigma_{\psi 2\mathrm{Cl}}(1,-2) = \left|\frac{1/8}{-3/8}\right| \sqrt{\frac{291}{128}} = 0.5026.$$

APPENDIX C

CALCULATION OF MOMENTS IN EXAMPLE 4.2

Consider the moments of the density $f_\theta(x) = \frac{1}{\sqrt{2\pi}\theta}e^{-\frac{(x-\theta)^2}{2\theta^2}}$. We calculate $\mu_k = E_\theta[X^k]$; $k = 1, 2, \ldots, 8$. Clearly, $E_\theta[X] = \mu_1 = \theta$. Then $E_\theta[X^2] = \sigma^2 + \mu^2 = \theta^2 + \theta^2 = 2\theta^2$. $E_\theta[X^3] = E_\theta[((X - \mu) + \mu)^3] = E_\theta[(X - \mu)^3 + 3\mu(X - \mu)^2 + 3\mu^2(X - \mu) + \mu^3] = 0 + 3\mu\sigma^2 + 0 + \mu^3 = 3\theta\theta^2 + \theta^3 = 4\theta^3$. Similarly, we find $E_\theta[X^4] = E_\theta[(X - \mu)^4 + 4(X - \mu)^3\mu + 6(X - \mu)^2\mu^2 + 4(X - \mu)\mu^3 + \mu^4]$. Note $E_\theta[(X - \mu)^4] = 3\sigma^4$. Hence, $E_\theta[X^4] = 3\sigma^4 + 6\mu^2\sigma^2 + \mu^4 = 3\theta^4 + 6\theta^4 + \theta^4 = 10\theta^4$. Now $E_\theta[X^5] = 5\mu E[(X - \mu)^4] + 10\mu^3 E[(X - \mu)^2] + \mu^5 = 5\mu 3\sigma^4 + 10\mu^3\sigma^2 + \mu^5 = 5\mu \times 3\sigma^4 + 10\mu^3\sigma^2 + \mu^5 = 26\theta^5$.

Now $E_\theta[X^6] = E_\theta[(X - \mu)^6 + 6(X - \mu)^5\mu + 15(X - \mu)^4\mu^2 + 20(X - \mu)^3\mu^3 + 15(X - \mu)^2\mu^4 + 6(X - \mu)\mu^5 + \mu^6]$. Also for a standard normal variable Z we have $E[Z^{2m}] = \frac{1}{\sqrt{\pi}}2^m\Gamma(m + \frac{1}{2})$, where $\Gamma(u)$ is the usual gamma function so that $E[Z^6] = \frac{1}{\sqrt{\pi}}2^3\Gamma(3 + \frac{1}{2}) = \frac{1}{\sqrt{\pi}}8\frac{5}{2}\frac{3}{2}\frac{1}{2}\Gamma(\frac{1}{2}) = 15$. Then it follows that $\mu_6 = 15\sigma^6 + 15 \times 3\sigma^4\mu^2 + 15\sigma^2\mu^4 + \mu^6 = 15\theta^6 + 45\theta^6 + 15\theta^6 + \theta^6 = 76\theta^6$.

To calculate $\mu_7 = E_\theta[X^7] = E_\theta[(X - \mu)^7 + 7(X - \mu)^6\mu + 21(X - \mu)^5\mu^2 + 35(X - \mu)^4\mu^3 + 35(X - \mu)^3\mu^4 + 21(X - \mu)^2\mu^5 + 7(X - \mu)\mu^6 + \mu^7]$, we see that $\mu_7 = 7 \times 15\sigma^6\mu + 0 + 35 \times 3\sigma^4\mu^3 + 0 + 21\sigma^2 + \mu^7 = 232\theta^7$.

Robustness Theory and Application, First Edition. Brenton R. Clarke.
© 2018 John Wiley & Sons, Inc. Published 2018 by John Wiley & Sons, Inc.
Companion website: www.wiley.com/go/clarke/robustnesstheoryandapplication

Finally, $\mu_8 = E_\theta[X^8] = E_\theta[(X-\mu)^8 + 8(X-\mu)^7\mu + 28(X-\mu)^6\mu^2 + 56(X-\mu)^5\mu^3 + 70(X-\mu)^4\mu^4 + 56(X-\mu)^3\mu^5 + 28(X-\mu)^2\mu^6 + 8(X-\mu)\mu^7 + \mu^8]$. Given $E[Z^8] = 105$ we find $\mu_8 = E_\theta[X^8] = 105\sigma^8 + 28 \times 15\sigma^6\mu^2 + 70 \times 3\sigma^4\mu^4 + 28\sigma^2\mu^6 + \mu^8 = 105\theta^8 + 420\theta^8 + 210\theta^8 + 28\theta^8 + \theta^8 = 764\theta^8$.

BIBLIOGRAPHY

Agostinelli, C., Marazzi, A., and Yohai, V.J. (2014). Robust estimators of the generalized log-gamma distribution. *Technometrics* **56**: 92–101.

Alexandroff, A.D. (1943). Additive set functions in abstract spaces. *Matematicheskii Sbornik. New Series* **13**: 169–234.

Amemiya, T. (1994). *Introduction to Statistics and Econometrics*. Cambridge, MA: Harvard University Press.

Anderson, T.W. and Darling, D.A. (1952). Asymptotic theory of certain "goodness of fit" criteria based on stochastic processes. *Annals of Mathematical Statistics* **23**: 193–212.

Andrews, D.F., Bickel, P.J., Hampel, F.R. et al. (1972). *Robust Estimates of Location: Survey and Advances*. Princeton: Princeton University Press.

Atkinson, A.C. (1981). Two graphical displays for outlying and influential observations in regression. *Biometrika* **68**: 13–20.

Atkinson, A.C. (1986a). Masking unmasked. *Biometrika* **73**: 533–541.

Atkinson, A.C. (1986b). Comment: aspects of diagnostic regression analysis. *Statistical Science* **1**: 397–401.

Atkinson, A.C. (1994). Fast very robust methods for the detection of multiple outliers. *Journal of the American Statististical Association* **89**: 1329–1339.

Atkinson, A.C. and Riani, M. (2000). *Robust Diagnostic Regression Analysis*. New York: Springer-Verlag.

Atkinson, A.C., Riani, M., and Cerioli, A. (2004). *Exploring Multivariate Data with the Forward Search*. New York: Springer-Verlag.

Bachmaier, M. (2007). Consistency of completely outlier-adjusted simultaneous redescending M-estimators of location and scale. *Advances in Statistical Analysis* **91**: 197–219.

Baker, M. (2016). Statisticians issue warning on P-values. *Nature* **531**: 151.

Ball, F.G. and Rice, J.A. (1992). Stochastic models for ion channels: introduction and bibliography. *Mathematical Biosciences* **112**: 189–206.

Ball, F.G., Cai, Y., Kadane, J.B., and O'Hagan, A. (1999). Bayesian inference for ion-channel gating mechanisms directly from single-channel recordings, using Markov chain Monte Carlo. *Proceedings of the Royal Society of London, A* **455**: 2879–2932.

Barnett, V.D. (1966). Evaluation of the maximum-likelihood estimator where the likelihood equation has multiple roots. *Biometrika* **53**: 151–165.

Barnett, V. and Lewis, T. (1994). *Outliers in Statistical Data*, 3e. New York: Wiley.

Basu, A. and Lindsay, B.G. (1994). Minimum disparity estimation for continuous models: efficiency, distributions and robustness. *Annals of the Institute of Statistical Mathematics* **46**: 683–705.

Basu, A., Harris, I.R., Hjort, N.L., and Jones, M.C. (1998). Robust and efficient estimation by minimizing a density power divergence. *Biometrika* **85**: 549–559.

Basu, A., Shioya, H., and Park, C. (2011). *Statistical Inference the Minimum Distance Approach*. New York: CRC Press.

Beaton, A.E. and Tukey, J.W. (1974). The fitting of power series meaning polynomials, illustrated on band-spectroscopic data. *Technometrics* **16**: 147–185.

Bednarski, T. (1985). On minimum bias and variance estimation for parametric models with shrinking contamination. *Probability and Mathematical Statistics* **6**: 121–129.

Bednarski, T. (1993a). Fréchet differentiability and robust estimation. In: *Asymptotic Statistics, Proceedings of the Fifth Prague Symposium 1994* (ed. P. Mandl and M. Hušková), 49–58. Berlin: Springer-Verlag.

Bednarski, T. (1993b). Robust estimation in Cox's regression model. *Scandinavian Journal of Statistics* **20**: 213–225.

Bednarski, T. (2004). Robust estimation in the generalized poisson model. *Statistics* **38**: 149–159.

Bednarski, T. (2010). Fréchet differentiability in statistical inference for time series. *Statistical Methods and Applications* **19**: 517–528.

Bednarski, T. and Clarke, B.R. (1993). Trimmed likelihood estimation of location and scale of the normal distribution. *Australian Journal of Statistics* **35**: 141–153.

Bednarski, T. and Clarke, B.R. (1998). On locally uniform expansions of regular functionals. *Discussiones Mathematicae, Algebra and Stochastic Methods* **18**: 155–165.

Bednarski, T. and Clarke, B.R. (2002). Asymptotics for an adaptive trimmed likelihood estimator. *Statistics* **36**: 1–8.

Bednarski, T., Clarke, B.R., and Kolkiewicz, W. (1991). Statistical expansions and locally uniform Fréchet differentiability. *Journal of the Australian Mathematical Society, Series A* **50**: 88–97.

Bednarski, T., Clarke, B.R., and Schubert, D. (2010). Adaptive trimmed likelihood in regression. *Discussiones Mathematicae Probability and Statistics* **30**: 203–219.

Beran, R. (1977). Minimum Hellinger distance estimates for parametric models. *The Annals of Statistics* **5**: 445–463.

Bernoulli, J. (1713). *Ars Conjectandi, Usum & Applicationem Praecedentis Doctrinae in Civilibus, Moralibus and Oeconomicis*. Basileae: Impensis Thurnisiorum, Fratrum.

Bickel, P.J. (1975). One-step Huber estimates in the linear model. *Journal of the American Statististical Association* **70**: 428–433.

Bickel, P.J. (1981). *Quelques Aspects de la Statistique Robuste*, Lecture Notes in Mathematics. Berlin: Springer.

Bickel, P.J. and Doksum, K.A. (2007). *Mathematical Statistics Basic Ideas and Selected Topics, Volume I*. Upper Saddle River, NJ: Pearson Prentice Hall.

Biernacki, C. (2005). Testing for a global maximum of the likelihood. *Journal of Computational and Graphical Statistics* **14**: 657–674.

Billingsley, P. (1956). The invariance principle for dependent random variables. *Transactions of the American Mathematical Society* **83**: 250–268.

Billingsley, P. (1968). *Convergence of Probability Measures*, 1e. New York: Wiley.

Billingsley, P. (2009). *Convergence of Probability Measures*, 2e. New York: Wiley.

Birch, J.B. (1980). Some convergence properties of iterated reweighted least squares in the location model. *Communications in Statistics: Simulation and Computation* **B9**: 359–369.

Blackman, J. (1955). On the approximation of a distribution function by an empiric distribution. *Annals of Mathematical Statistics* **26**: 256–267.

Blatt, D. and Hero, A.O. (2007). On tests for global maximum of the log-likelihood function. *IEEE Transactions on Information Theory* **53**: 2510–2525.

Bölthausen, E. (1977). Convergence in distribution of minimum-distance estimators. *Metrika* **24**: 215–227.

Boos, D.D. (1979). A differential for L-statistics. *The Annals of Statistics* **7**: 955–959.

Boos, D.D. (1981). Minimum distance estimators for location and goodness of fit. *Journal of the American Statistical Association* **76**: 663–670.

Boos, D.D. and Serfling, R.J. (1980). A note on differentials and the CLT and LIL for statistical functions, with application to M-estimates. *The Annals of Statistics* **8**: 618–624.

Bowman, K.O. and Shenton, L.R. (1988). *Properties of Estimators for the Gamma Distribution*. New York: Marcel Dekker Inc.

Box, G.E.P. (1953). Non normality and tests for variances. *Biometrika* **40**: 318–335.

Broffitt, B., Clarke, W.R., and Lachenbruch, P.A. (1980). The effect of Huberizing and trimming on the quadratic discriminant function. *Communications in Statistics: Theory and Methods* **9**: 13–25.

Brownlee, K.A. (1965). *Statistical Theory and Methodology in Science and Engineering*, 2e. New York: Wiley.

Butler, R.W. (1982). Nonparametric interval and point prediction using data trimmed by a Grubbs-type outlier rule. *The Annals of Statistics* **10**: 197–204.

Butler, R.W., Davies, P.L., and Jhun, M. (1993). Asymptotics for the minimum covariance determinant estimator. *The Annals of Statistics* **21**: 1385–1400.

Cai, J. (2004). Mixtures of exponential distributions. In: *Encyclopedia of Actuarial Science*, vol. **2** (ed. J.L. Teugels and B. Sundt), 1128–1131. Wiley.

Campbell, N.A. (1980). Robust procedures in multivariate analysis I: robust covariance estimation. *Applied Statistics* **29**: 231–237.

Campbell, N. (2016). S-estimation for some bivariate data, CSIRO Data61, Leeuwin Centre, 65 Brockway Road Floreat WA 6014 Research Report.

Carroll, R.J. and Ruppert, D. (1988). *Transformations and Weighting in Regression*. London: Chapman and Hall.

Casella, G. and Berger, R.L. (2002). *Statistical Inference*. Pacific Grove: Duxbury.

Cator, E.A. and Lopuhaä, H.P. (2012). Central limit theorem and influence function for the MCD estimators at general multivariate distributions. *Bernoulli* **18**: 520–551.

Cerioli, A., Riani, M., Atkinson, A.C., and Corbellini, A. (2017). The power of monitoring: how to make the most of a contaminated multivariate sample. Statistical Methods & Applications submitted.

Chambers, R.L. and Heathcote, C.R. (1981). On the estimation of slope and the identification of outliers in linear regression. *Biometrika* **68**: 21–33.

Choi, K. and Bulgren, W.G. (1968). An estimation procedure for mixtures of distributions. *Journal of the Royal Statistics Society, Series B* **30**: 444–460.

Christmann, A. and Van Messem, A. (2008). Bouligand derivatives and robustness of support vector machines for regression. *Journal of Machine Learning Research* **9**: 915–936.

Chung, K.L. (2001). *A Course in Probability Theory*, 3e. New York: Academic Press.

Clarke, B.R. (1983a). Uniqueness and Fréchet differentiability of functional solutions to maximum likelihood equations. *The Annals of Statistics* **11**: 1196–1205.

Clarke, F.H. (1983b). *Optimization and Nonsmooth Analysis*. New York: Wiley.

Clarke, B.R. (1986a). Asymptotic theory for description of regions in which Newton-Raphson iterations converge to location M-estimators. *Journal of Statistical Planning and Inference* **15**: 71–85.

Clarke, B.R. (1986b). Nonsmooth analysis and Fréchet differentiability of M-functionals. *Probability Theory and Related Fields* **73**: 197–209.

Clarke, B.R. (1989). An unbiased minimum distance estimator of the proportion parameter in a mixture of two normal distributions. *Statistics and Probability Letters* **7**: 275–281.

Clarke, B.R. (1990). The selection functional. *Probability and Mathematical Statistics* **11** (Fasc. 2): 149–156.

Clarke, B.R. (1994). Empirical evidence for adaptive confidence intervals and identification of outliers using methods of trimming. *Australian Journal of Statistics* **36**: 45–58.

Clarke, B.R. (1997). Testing for outliers in gold assay data; unweighted and weighted samples. In: *Proceedings of Workshop on Statistical Applications in Mining 1 October 1996* (ed. B.R. Clarke and I.W. Wright), 90–103. Murdoch, WA: Murdoch University Press.

Clarke, B.R. (2000a). A remark on robustness and weak continuity of M-estimators. *Journal of the Australian Mathematical Society (Series A)* **68**: 411–418.

Clarke, B.R. (2000b). A review of differentiability in relation to robustness with an application to seismic data analysis. *Proceedings of the Indian National Science Academy Part A* **66**: 467–482.

Clarke, B.R. (2000c). An adaptive method of estimation and outlier detection applicable for small to medium sample sizes. *Discussiones Mathematicae* **20**: 25–50.

Clarke, B.R. (2008). *Linear Models the Theory and Application of Analysis of Variance.* Hoboken, NJ: Wiley.

Clarke, B.R. (2014). Book review: methodology in robust and nonparametric statistics, by Jurečková, J. Sen, P. K. and Picek, J. CRC Press, 2012. *Australian & New Zealand Journal of Statistics* **56**: 497–499.

Clarke, B.R. and Futschik, A. (2007). On the convergence of Newton's method when estimating higher dimensional parameters. *Journal of Multivariate Analysis* **98**: 916–931.

Clarke, B.R. and Heathcote, C.R. (1978). Comment on "Estimating mixtures of normal distributions and switching regressions" by R.E.Quandt, and J.B. Ramsey. *Journal of the American Statististical Association* **73**: 749–750.

Clarke, B.R. and Heathcote, C.R. (1994). Robust estimation of k-component univariate normal mixtures. *Annals of the Institute of Statistical Mathematics* **46**: 83–93.

Clarke, B.R. and Lewis, T. (1998). An outlier problem in determination of ore grade. *Journal of Applied Statistics* **25**: 751–762.

Clarke, B.R. and McKinnon, P.L. (2005). Robust inference and modelling for the single ion channel. *Journal of Statistical Computation and Simulation* **75**: 513–529.

Clarke, B.R. and Milne, C.J. (2004). Small sample bias correction for Huber's Proposal-2 scale M-estimator. *Australian & New Zealand Journal of Statistics* **46**: 649–656.

Clarke, B.R. and Milne, C.J. (2013). A small sample bias correction and implications for inference. *Proceedings of 59th ISI World Statistics Congress*, Hong Kong (25–30 August 2013).

Clarke, B.R. and Schubert, D.D. (2006). An adaptive trimmed likelihood algorithm for identification of multivariate outliers. *Australian & New Zealand Journal of Statistics* **48**: 353–371.

Clarke, B.R., Yeo, G.F., and Milne, R.K. (1993). Local asymptotic theory for multiple solutions of likelihood equations, with application to a single ion channel model. *Scandinavian Journal of Statistics* **20**: 133–146.

Clarke, B.R., Gamble, D.K., and Bednarski, T. (2000). A note on the robustness of the β-trimmed mean. *Australian & New Zealand Journal of Statistics* **42**: 113–117.

Clarke, B.R., Gamble, D.K., and Bednarski, T. (2011). On a note on robustness of the β-trimmed mean. *Australian & New Zealand Journal of Statistics* **53**: 481.

Clarke, B.R., McKinnon, P.L., and Riley, G. (2012). A fast robust method for fitting gamma distributions. *Statistical Papers* **53**: 1001–1014.

Clarke, B.R., Davidson, T., and Hammarstrand, D. (2017a). A comparison of the L_2 minimum distance estimator and the EM-algorithm when fitting k-component univariate normal mixtures. *Statistical Papers* **58**: 1247–1266.

Clarke, B.R., Höller, A., Müller, C.H., and Wamahiu, K. (2017b). Investigation of the performance of trimmed estimators of the lifetime distributions with censoring. *Australian & New Zealand Journal of Statistics* **59**: 513–525.

Colquhoun, D. and Hawkes, A.D. (1982). On the stochastic properties of bursts of single ion channel openings and of clusters of bursts. *Philosophical Transactions. Royal Society of London B* **300**: 1–59.

Copas, J.B. (1975). On the unimodality of the likelihood for the Cauchy distribution. *Biometrika* **62**: 701–704.

Cox, D.R. and Hinkley, D.V. (1974). *Theoretical Statistics*. London: Chapman and Hall.

Cramér, H. (1928). On the composition of elementary errors. *Scandinavian Actuarial Journal* **1**: 13–74.

Cramér, H. (1946). *Mathematical Methods in Statistics*. Princeton, NJ: Princeton University Press.

Cressie, N. (1980). Relaxing assumptions in the one sample t-test. *Australian Journal of Statistics* **22**: 143–153.

Cressie, N. and Read, T.R.C. (1984). Multinomial goodness-of-fit tests. *Journal of the Royal Statistics Society, Series B* **46**: 440–464.

Cushny, A.R. and Peebles, A.R. (1905). The action of optical isomers. II. Hyoscines. *The Journal of Physiology* **32**: 501–510.

Cutler, A. and Cordiero-Braña, O.I. (1996). Minimum Hellinger distance estimation for finite mixture models. *Journal of the American Statistical Association* **91**: 1716–1723.

D'Agostino, R.B. and Stephens, M.A. (1986). *Goodness-of-Fit Techniques*. New York: Marcel Dekker.

Darling, D.A. (1955). The Cramér-Smirnov test in the parametric case. *Annals of Mathematical Statistics* **26**: 1–20.

Davies, L. (1992a). An efficient Fréchet differentiable high breakdown multivariate location and dispersion estimator. *Journal of Multivariate Analysis* **40**: 311–327.

Davies, L. (1992b). The asymptotics of Rousseeuw's minimum volume ellipsoid estimator. *The Annals of Statistics* **20**: 1828–1843.

Davison, A.C. and Hinkley, D.V. (1997). *Bootstrap Methods and their Application*. Cambridge: Cambridge University Press.

Diaconis, P., Holmes, S., and Montgomery, R. (2007). Dynamical bias in the coin toss. *SIAM Review* **49**: 211–235.

Donoho, D.L. and Huber, P.J. (1983). The notion of breakdown point. In: *A Festscrift for Eric L. Lehmann* (ed. P. Bickel, K. Doksum and J.L. Hodges Jr.), 157–184. Belmont, CA: Wadsworth.

Donoho, D.L. and Liu, R.C. (1988a). The "automatic" robustness of minimum distance functionals. *The Annals of Statistics* **16**: 552–586.

Donoho, D.L. and Liu, R.C. (1988b). Pathologies of some minimum distance estimators. *The Annals of Statistics* **16**: 587–608.

Draper, N.R. and Smith, H. (1966). *Applied Regression Analysis*. New York: Wiley.

Durbin, J. and Knott, M. (1972). Components of the Cramér–von Mises statistic. *Journal of the Royal Statistics Society, Series B* **34**: 290–307.

Efron, B. (1979). Bootstrap methods: another look at the jackknife. *The Annals of Statistics* **7**: 1–26.

Efron, B. (1982). *The Jackknife, the Bootstrap and Other Resampling Plans*. Philadelphia, PA: Society for Industrial and Applied Mathematics.

Efron, B. and Tibshirani, R.J. (1993). *An Introduction to the Bootstrap*. New York: Chapman and Hall.

Elker, J., Pollard, D., and Stute, W. (1979). Glivenko-Cantelli theorems. *Advances in Applied Probability* **11**: 820–833.

Everitt, B.S. and Hand, D.J. (1981). *Finite Mixture Distributions*. London: Chapman and Hall.

Eyer, S. and Riley, G. (1999). Measurement quality assurance in a production system for bauxite analysis by FTIR. North American Chapter of the International Chemometrics Society, Newsl No. 19.

Fasano, M.V., Maronna, R.A., Sued, M., and Yohai, V.J. (2012). Continuity and differentiability of regression M functionals. *Bernoulli* **18**: 1284–1309.

Feigin, P.D. and Heathcote, C.R. (1976). The empirical characteristic function and the Cramér–von Mises statistic. *Sankhyā, Series A* **38**: 309–325.

Feller, W. (1966). *An Introduction to Probability Theory and Its Applications, Volume II*. New York: Wiley.

Fernholz, L. (1983). *Von Mises Calculus for Statistical Functionals VIII*. New York: Springer-Verlag.

Finger, R. (2010). Review of "robustbase" software for R. *Journal of Applied Econometrics* **25**: 1205–1210.

Fisher, R.A. (1925). Theory of statistical estimation. *Proceedings of the Cambridge Philological Society* **22**: 700–725.

Foutz, R.V. (1977). On the unique consistent solutions to the likelihood equations. *Journal of the American Statistical Association* **72**: 147–148.

Foutz, R. and Srivastava, R. (1979). Statistical inference for Markov processes when the model is incorrect. *Advances in Applied Probability* **11**: 737–749.

Fréchet, M. (1925). La notion de différentielle dans l'analyse générale. *Annales Scientifiques de l'École Normale Supérieure* **42**: 293–323.

Freedman, D. and Diaconis, P. (1982). On inconsistent M-estimators. *The Annals of Statistics* **10**: 454–461.

Freue, G.V.C. (2007). The Pitman estimator of the Cauchy location parameter. *Journal Statistical Planning and Inference* **137**: 1900–1913.

Gallegos, M.T. and Ritter, G. (2009). Trimmed ML estimation of contaminated mixtures. *Sankhya, Series A* **71**: 164–220.

Gan, L. and Jiang, J. (1999). A test for global maximum. *Journal of the American Statistical Association* **94**: 847–854.

Garcia-Pérez, A. (2016). A von Mises approximation to the small sample distribution of the trimmed mean. *Metrika* **79**: 369–388.

Ghosh, A. (2014). Asymptotic properties of minimum S-divergence estimator for discrete models. *Sankhyā, A* **77**: 380–407.

Ghosh, A. and Basu, A. (2017). The minimum S-divergence estimator under continuous models: the Basu-Lindsay approach. *Statistical Papers* **58**: 341–372.

Gill, R.D. and Heesterman, C.C. (1992). A central limit theorem for M-estimators by the von Mises method. *Statistica Neerlandica* **46**: 165–177.

Gill, R.D., Wellner, J.A., and Praestgaard, J. (1989). Non-and semi-parametric maximum likelihood estimators and the von Mises method(part i) [with discussion and reply]. *Scandinavian Journal of Statistics* **16**: 97–128.

Hadi, A.S. (1992). Identifying outliers in multivariate data. *Journal of the Royal Statistics Society, Series B* **52**: 761–771.

Hadi, A.S. (1994). A modification of a method for the detection of outliers in multivariate samples. *Journal of the Royal Statistics Society, Series B* **56**: 393–396.

Hadi, A.S. and Simonoff, J.S. (1993). Procedures for the identification of multiple outliers in linear models. *Journal of the American Statistical Association* **88**: 1264–1272.

Hall, P.G. (1992). *The Bootstrap and Edgeworth Expansion*. New York: Springer-Verlag.

Hampel, F. R. (1968). Contributions to the theory of robust estimation. PhD thesis. University of California, Berkeley.

Hampel, F.R. (1971). A general qualitative definition of robustness. *Annals of Mathematical Statistics* **42**: 1887–1896.

Hampel, F.R. (1974). The influence curve and its role in robust estimation. *Journal of the American Statististical Association* **69**: 383–393.

Hampel, F.R. (1978). Optimally bounding the gross-error-sensitivity and the influence of position in factor space. *Proceedings of the ASA Statistical Computing Section, ASA*, Washington, DC, pp. 59–64.

Hampel, F.R. (1992). Introduction to Huber (1964) robust estimation of a location parameter. In: *Breakthroughs in Statistics 2. Methodology and Distribution* (ed. S. Kotz and N.L. Johnson), 479–491. New York: Springer.

Hampel, F.R., Ronchetti, E.M., Rosseeuw, P.J., and Stahel, W.A. (1986). *Robust Statistics, the Approach Based on Influence Functions*. New York: Wiley.

Heathcote, C.R. (1977). The integrated squared error estimation of parameters. *Biometrika* **64**: 255–264.

Heathcote, C.R. and Silvapulle, M.J. (1981). Minimum mean squared estimation of location and scale parameters under misspecification of the model. *Biometrika* **68**: 501–514.

Heritier, S. and Ronchetti, E. (1994). Robust bounded influence tests in general parametric models. *Journal of the American Statistical Association* **89**: 897–904.

Heritier, S., Cantoni, E., Copt, S., and Victoria-Feser, M.P. (2009). *Robust Methods in Biostatistics*. New York: Wiley.

Hettmansperger, T.P., Heuter, I., and Hüsler, J. (1994). Minimum distance estimators. *Journal of Statistical Planning and Inference* **41**: 291–302.

Heyde, C.C. (1997). *Quasi-Likelihood and Its Application*. New York: Springer-Verlag.

Heyde, C.C. and Morton, R. (1998). Multiple roots on general estimating equations. *Biometrika* **85**: 954–959.

Hill, R. W. (1977). Robust regression when there are outliers in the carriers. Unpublished PhD dissertation. Harvard University, Cambridge, MA.

Huber, P.J. (1964). Robust estimation of a location parameter. *Annals of Mathematical Statistics* **35**: 73–101.

Huber, P.J. (1967). The behaviour of maximum likelihood estimates under non-standard conditions. *Proceedings of the Fifth Berkeley Symposium on Mathematical Statistics and Probability 1*, 221–223. Berkeley, CA: University of California Press. https://projecteuclid.org/euclid.bsmsp/1200512988.

Huber, P.J. (1973). Robust regression: asymptotics, conjectures and Monte Carlo. *The Annals of Statistics* **1**: 799–821.

Huber, P.J. (1977). *Robust Statistical Procedures*, 1e. Philadelphia, PA: Society for Industrial and Applied Mathematics.

Huber, P.J. (1981). *Robust Statistics*, 1e. New York: Wiley.

Huber, P.J. (1996). *Robust Statistical Procedures*. 2e, CBMS-NSF Regional Conference Series in Applied Mathematics. Philadelphia, PA: Society for Industrial and Applied Mathematics.

Huber, P.J. and Ronchetti, E.M. (2009). *Robust Statistics*, 2e. New York: Wiley.

Huber-Carol, C. (1970). Etude asymptotique de tests robustes. PhD thesis. ETH, Zürich.

Hubert, M. and Debruyne, M. (2009). Minimum covariance determinant. *Wiley Interdisciplinary Reviews: Computational Statistics* **2**: 36–43.

Hubert, M., Rousseeuw, P.J., and van Aelst, S. (2008). High-breakdown robust multivariate methods. *Statistical Science* **23**: 92–119.

Hubert, M., Rousseeuw, P.J., and Verdonck, T. (2012). A deterministic algorithm for robust location and scatter. *Journal of Computational and Graphical Statistics* **21**: 618–637.

Huzurbazaar, V.S. (1948). The likelihood equation, consistency and the maxima of the likelihood equation. *Annals of Eugenics* **14**: 185–200.

Jaeckel, L.A. (1971a). Robust estimates of location: symmetric and asymmetric contamination. *Annals of Mathematical Statistics* **42**: 1020–1034.

Jaeckel, L.A. (1971b). Some flexible estimates of location. *Annals of Mathematical Statistics* **42**: 1540–1552.

Jewell, N.P. (1982). Mixtures of exponential distributions. *Annals of Statistics* **10**: 479–484.

Jurečková, J., Sen, P.K., and Picek, J. (2012). *Methodology in Robust and Nonparametric Statistics*. Boca Raton, FL: CRC Press.

Kac, M., Kiefer, J., and Wolfowitz, J. (1955). On tests of normality and other tests of goodness of fit based on distance methods. *Annals of Mathematical Statistics* **26**: 189–211.

Kallianpur, G. (1963). Von Mises functionals and maximum likelihood estimation. *Sankhyā Series A* **23**: 149–158.

Kallianpur, G. and Rao, C.R. (1955). On Fisher's lower bound to asymptotic variance of a consistent estimate. *Sankhyā* **15**: 331–342.

Kianifard, F. and Swallow, W.H. (1989). Using recursive residuals, calculated on adaptively-ordered observations, to identify outliers in linear regression. *Biometrics* **45**: 571–585.

Kianifard, F. and Swallow, W.H. (1990). A Monte Carlo comparison of five procedures for identifying outliers in linear regression. *Communications in Statistics, Part A – Theory and Methods* **19**: 1913–1938.

Kiefer, J. (1961). On large deviations of the empiric D.F. of vector chance variables and a law of iterated logarithm. *Pacific Journal of Mathematics* **11**: 649–660.

Kiesel, R., Rühlicke, R., Stahl, G., and Zheng, J. (2016). The Wasserstein metric and robustness in risk management. *Risks* **4** (3): 32.

Knüsel, L.F. (1969). Über minimum-distance-Schätzungen. PhD thesis. ETH, Zürich.

Kohl, M. and Ruckdeschel, P. (2010). R package distrMod: S4 classes and methods for probability models. *Journal of Statistical Software* **35**: 1–27.

Kolmogorov, A.N. (1933). Sulla determinazione empirica di una legge di distribuzione. *Giornale dell'Instituto Italiano degli Attuari* **4**: 83–91.

Koul, H.L. (1992). *Weighted Empiricals and Linear Models*, IMS Lecture Notes Monograph Series, vol. **21**. Hayward, CA: Institute of Mathematical Statistics.

Koul, H. and DeWet, T. (1983). Minimum distance estimation in a linear regression model. *The Annals of Statistics* **11**: 921–932.

Kozek, A. (1998). On minimum distance estimation using Kolmogorov-Lévy type metrics. *Australian & New Zealand Journal of Statistics* **40**: 317–333.

Kulawik, A. and Zontek, S. (2015). Robust estimation in the multivariate normal model with variance components. *Statistics* **49**: 766–780.

Le Cam, L. (1953). On some asymptotic properties of maximum likelihood estimates and related Bayes' estimates. *University of California Publications in Statistics* **1**: 277–330.

Leitch, R.A. and Paulson, A.S. (1975). Estimation of stable law parameters: stock price application. *Journal of the American Statistical Association* **70**: 690–697.

Lindsay, B.G. (1994). Efficiency versus robustness: the case for minimum Hellinger distance and related methods. *The Annals of Statistics* **22**: 1081–1114.

Lohse, K. (1987). The consistency of the bootstrap. *Statistics and Decisions* **5**: 353–366.

Lukacs, E. (1970). *Characteristic Functions*. Darien, CN: Hafner Publishing Co.

Luong, A. (2016). Cramér–von Mises distance estimation for some positive infinitely divisible parametric families with actuarial applications. *Scandinavian Actuarial Journal* **6**: 530–549.

Macdonald, P.D.M. (1971). Comment on "An estimation procedure for mixtures of distributions." by K. Choi, and W.G. Bulgren. *Journal of the Royal Statistics Society, Series B* **33**: 326–329.

Marazzi, A. and Ruffieux, C. (1996). Implementing M-estimators of the gamma distribution. In: *Robust Statistics, Data Snalysis, and Computer Intensive Methods in Honor of Peter Hubers 60th Birthday*, Lecture Notes in Statistics (ed. H. Rieder), 277–297. Heidelberg: Springer.

Markatou, M., Chen, Y., Afrendras, G., and Lindsay, B.G. (2016). Statistical distances and their role in robustness. Personal Communication.

Maronna, R.A. and Yohai, V.J. (1981). Asymptotic behaviour of general M-estimates for regression and scale with random carriers. *Zeitschrift für Wahrscheinlichkeitstheorie und Verwandte Gebiete* **58**: 7–20.

Maronna, R.A., Martin, D.R., and Yohai, V.J. (2006). *Robust Statistics: Theory and Methods*. New York: Wiley.

Martin, E.S. (1936). A study of an Egyptian series of mandibles. With special reference to mathematical methods of sexing. *Biometrika* **28**: 149–178.

Matusita, K. (1953). On the estimation by the minimum distance method. *Annals of the Institute of Statistical Mathematics* **5**: 57–65.

Matusita, K. (1955). Decision rules, based on the distance, for problems of fit, two samples, and estimation. *Annals of Mathematical Statistics* **26**: 631–640.

McLachlan, G.J. and Basford, K.E. (1988). *Mixture Models: Inference and Applications to Clustering*. New York: Marcel Dekker.

McLachlan, G.J. and Peel, D. (2000). *Finite Mixture Models*, 1e. New York: Wiley.

Millar, P.W. (1981). Robust estimation via minimum distance methods. *Zeitschrift für Wahrscheinlichkeitstheorie und Verwandte Gebiete* **55**: 72–89.

Millar, P.W. (1982). Optimal estimation of a general regression function. *The Annals of Statistics* **10**: 717–740.

von Mises, R.E. (1928). *Wahrscheinlichkeit, Statistik Und Wahrheit*. Vienna: Julius Springer.

von Mises, R. (1947). On the asymptotic distribution of differentiable statistical functions. *Annals of Mathematical Statistics* **18**: 309–348.

Mizera, I. (1996). Weak continuity of redescending M-estimators of location with unbounded objective function. In: *PROBSTAT '94. Proceedings of the Second International Conference on Mathematical Statistics*, vol. 7 (ed. A. Pázman and V. Witkovský), 343–347. Tatra Mountains Mathematical Publication.

Mlodinow, L. (2008). *The Drunkard's Walk: How Randomness Rules Our Lives*. New York: Random House.

Müller, C.H. (1997). *Robust Planning and Analysis of Experiments*. New York: Springer-Verlag.

Müller, C.H., Szugat, S., Celik, N., and Clarke, B.R. (2016). Influence functions of trimmed likelihood estimators for lifetime experiments. *Statistics* **50**: 505–524.

Neykov, N. and Müller, C.H. (2003). Breakdown point and computation of trimmed likelihood estimators in generalized linear models. In: *Developments in Robust Statistics* (ed. R. Dutter, P. Filzmoser, U. Gather and P.J. Rousseeuw), 277–286. Heidelberg: Physica-Verlag.

Neykov, N.M. and Neytchev, P.N. (1990). A robust alternative of the ML estimators. *Proceedings of COMPSTAT Short Communications, Dubrovnik, Yugoslavia '90*, 99–100.

Neykov, N., Filzmoser, P., Dimova, R., and Neytchev, P. (2007). Robust fitting of mixtures using the trimmed likelihood estimator. *Computational Statistics and Data Analysis* **52**: 299–308.

O'Gorman, T.W. (2012). *Adaptive Tests of Significance Using Permutations of Residuals with R and SAS*. Hoboken, NJ: Wiley.

Omelka, M. and Salibián-Barrera, M. (2010). Uniform asymptotics for S- and MM-regression estimators. *Annals of the Institute of Statistical Mathematics* **62**: 897–927.

Öztürk, O. and Hettmansperger, T.P. (1997). Generalised weighted Cramér–von Mises distance estimators. *Biometrika* **84**: 283–294.

Parr, W.C. (1985). Jackknifing differentiable statistical functionals. *Journal of the Royal Statistics Society, Series B* **47**: 56–66.

Parr, W.C. and DeWet, T. (1981). On minimum Cramér–von Mises-norm parameter estimation. *Communication in Statistics* **A10**: 1149–1166.

Parr, W.C. and Schucany, W.R. (1980). Minimum distance and robust estimation. *Journal of the American Statististical Association* **75**: 616–624.

Patra, S., Maji, A., and Basu, A. (2013). The power divergence and the density power divergence families: the mathematical connection. *Sankhyā, B* **75**: 16–28.

Paulson, A.S., Holcomb, E.W., and Leitch, R.A. (1975). The estimation of the parameters of the stable laws. *Biometrika* **62**: 163–170.

Pollard, D. (1980). The minimum distance method of testing. *Metrika* **27**: 43–70.

Pollard, D. (1990). *Empirical Processes: Theory and Applications*, NSF-CBMS Regional Conference Series in Probability and Statistics, vol. **2**. Hayward, CA and Alexandria, VA: Institute of Mathematical Statistics and the American Statistical Association.

Prohorov, Y.V. (1956). Convergence of random processes and limit theorems in probability. *Theory of Probability and Its Applications* **1**: 157–214.

Quandt, R.E. (1972). A new approach to switching regressions. *Journal of the American Statististical Association* **67**: 306–310.

Quandt, R.E. and Ramsey, J.B. (1978). Estimating mixtures of normal distributions and switching regressions. *Journal of the American Statististical Association* **73**: 730–738.

Quenouille, M.H. (1949). Problems in plane sampling. *Annals of Mathematical Statistics* **20**: 355–375.

Quenouille, M.H. (1956). Notes on bias in estimation. *Biometrika* **43**: 353–360.

Rao, C.R. (1957). Maximum likelihood estimation for the multinomial distribution. *Sankhyā* **18**: 139–148.

Rao, R.R. (1962). Relations between weak and uniform convergence convergence of measures with applications. *Annals of Mathematical Statistics* **33**: 659–680.

Rao, C.R. (1973). *Linear Statistical Inference and Its Applications*. New York: Wiley.

Reeds, J.A. (1976). On the definition of von Mises functionals. PhD thesis, Harvard University, Cambridge, MA.

Reeds, J.A. (1985). Asymptotic number of roots of Cauchy location likelihood equations. *The Annals of Statistics* **13**: 775–784.

Rey, W.J.J. (1977). M-estimators in robust regression, a case study. In: *Recent Developments in Statistics* (ed. J.R. Barra, F. Brodeau, G. Romier and B. Van Custom), 591–594. North Holland: Elsevier.

Rieder, J.A. (1978). A robust asymptotic testing model. *The Annals of Statistics* **6**: 1080–1094.

Rieder, H. (1994). *Robust Asymptotic Statistics*. New York: Springer-Verlag.

Rivest, L. (1982). Some asymptotic distributions in the location scale model. *Annals of the Institute of Statistical Mathematics* **34**: 225–239.

Rocke, M.R. and Woodruff, D.L. (1996). Identification of outliers in multivariate data. *Journal of the American Statistical Association* **91**: 1047–1061.

Rousseeuw, P.J. (1983). Multivariate estimation with high breakdown point. In: *Mathematical Statistics and Applications*. (1985), vol. **B** (ed. W. Grossman, G. Pflug, I. Vincze and W. Wertz), 283–297. Dordrecht: Reidel Publishing Co.

Rousseeuw, P.J. (1984). Least median of squares regression. *Journal of the American Statistical Association* **79**: 871–880.

Rousseeuw, P.J. and Leroy, A.M. (1987). *Robust Estimation and Outlier Detection*. New York: Wiley.

Rousseeuw, P.J. and Ronchetti, E. (1979). *The Influence Curve for Tests*, Research Report/Fachgruppe für Statistik, vol. **21**. Zürich: ETH.

Rousseeuw, P.J. and Ronchetti, E. (1981). Influence curves for general statistics. *Journal of Computational and Applied Mathematics* **7**: 161–166.

Rousseeuw, P.J. and Yohai, V.J. (1984). Robust regression by means of S-estimators. In: *Robust and Nonlinear Time Series Analysis*, Lecture Notes in Statistics (ed. J. Franke, W. Härdle and R.D. Martin), 256–272. New York: Springer-Verlag.

Sahler, W. (1970). Estimation by minimum-discrepancy methods. *Metrika* **16**: 85–106.

Sakmann, B. and Neher, E. ed. (1995). *Single-Channel Recording*, 2e. New York: Plenum.

Salibian-Barrera, M. and Yohai, V.J. (2012). A fast algorithm for S-regression estimates. *Journal of Computational and Graphical Statistics* **15**: 414–427.

Salibian-Barrera, M. and Zamar, R.H. (2002). Bootstrapping robust estimates of regression. *The Annals of Statistics* **30**: 556–582.

Salibian-Barrera, M. and Zamar, R.H. (2004). Uniform asymptotics for robust location estimates when the scale is unknown. *The Annals of Statistics* **32**: 1432–1447.

Salibian-Barrera, M., Willems, G., and Zamar, R. (2008). The fast-τ estimator for regression. *Journal of Computational and Graphical Statistics* **17**: 659–672.

Schubert, D. D. (2005). A multivariate adaptive trimmed likelihood algorithm. PhD thesis. Murdoch University, Murdoch, WA.

Scott, D.W. (2001). Parametric statistical modeling by minimum integrated squared error. *Technometrics* **43**: 274–285.

Serfling, R.J. (1980). *Approximation Theorems of Mathematical Statistics*. New York: Wiley.

Shao, J. (1989). The efficiency and consistency of approximations to the jackknife variance estimators. *Journal of the American Statistical Association* **84**: 114–119.

Shao, J. (1993). Differentiability of statistical functionals and consistency of the jackknife. *The Annals of Statistics* **21**: 61–75.

Shao, J. and Tu, D. (1995). *The Jackknife and Bootstrap*. New York: Springer.

Shao, J. and Wu, C.F.J. (1989). A general theory for jackknife variance estimation. *The Annals of Statistics* **17**: 1176–1197.

Sigg, D. (2014). Modeling ion channels: past, present, and future. *The Journal of General Physiology* **144**: 7–26.

Silvapulle, M.J. (1981). The minimum omega squared-method of estimation. PhD thesis. Australian National University, Canberra.

Siverman, B.W. (1986). *Density Estimation for Statistics and Data Analysis*. London: Chapman and Hall.

Small, C.G. (2010). *Expansions and Asymptotics for Statistics*. Boca Raton, FL: Chapman and Hall/CRC.

Small, C.G. and Wang, J. (2003). *Numerical Methods for Nonlinear Estimating Equations*. Oxford: Clarendon Press.

Small, C.G. and Yang, Z. (1999). Multiple roots of estimating functions. *The Canadian Journal of Statistics* **27**: 585–598.

Small, C.G., Wang, J., and Yang, Z. (2000). Eliminating multiple root problems in estimation. *Statistical Science* **15**: 313–341.

Smirnov, N.V. (1939). Estimation of deviation between empirical distribution functions in two independent samples (in Russian). *Bulletin of Moscow University* **2**: 3–16.

Staudte, R.G. and Sheather, S.J. (1990). *Robust Estimation and Testing*. New York: Wiley.

Stigler, S.M. (1977). Do robust estimators work with real data. *The Annals of Statistics* **5**: 1055–1098.

Stigler, S.M. (1999). *Statistics on the Table the History of Statistical Concepts and Methods*. Cambridge, MA: Harvard University Press.

Tarone, R.E. and Gruenage, G. (1975). A note on the uniqueness of roots of the likelihood equations for vector-valued parameters. *Journal of the American Statistical Association* **70**: 903–904.

Thieler, A.M., Fried, R., and Rathjens, J. (2016). RobPer: an R package to calculate periodograms for light curves based on robust regression. *Journal of Statistical Software* **69**: 1–36.

Thornton, J.C. and Paulson, A.S. (1977). Asymptotic distribution of characteristic function-based estimators for the stable laws. *Sankhyā, Series A* **39**: 341–354.

Titterington, D.M., Smith, A.F.M., and Makov, U.E. (1985). *Statistical Analysis of Finite Mixture Distributions*. New York: Wiley.

Topsoe, F. (1970). On the Glivenko-Cantelli theorem. *Zeitschrift für Wahrscheinlichkeitstheorie und Verwandte Gebiete*. **14**: 239–250.

Tukey, J.W. (1958). Bias and confidence in not-quite large samples (preliminary report). *Annals of Mathematical Statistics* **29**: 614.

Tukey, J.W. (1960). A survey of sampling from contaminated distributions. In: *Contributions to Probability and Statistics* (ed. I. Olkin), 448–485. Stanford, CA: Stanford University Press.

Tukey, J.W. and McLaughlin, D.H. (1963). Less vulnerable confidence and significance procedures for location based on a single sample: trimming/Winsorization 1. *Sankhyā Series A* **25**: 331–352.

van der Vaart, A.W. (1991). Efficiency and Hadamard differentiability. *Scandinavian Journal of Statistics* **18**: 63–75.

van der Vaart, A.W. and Wellner, J.A. (1996). *Weak Convergence and Empirical Processes, with Applications to Statistics*. New York: Springer.

Vandev, D.L. and Neykov, N.M. (1993). Robust maximum likelihood in the Gaussian case. In: *New Directions in Statistical Data Analysis and Robustness* (ed. E. Ronchetti and W.A. Stahel), 259–264. Basel: Birkhäuser.

Vapnik, V. and Chervonenkis, A. (2004). On the uniform convergence of relative frequencies of events to their probabilities. *Theory of Probability and Its Applications* **16**: 264–280.

Varadarajan, V.S. (1958). On the convergence of probability distributions. *Sankhyā* **19**: 23–26.

Vecchia, D.L., Ronchetti, E., and Trojani, F. (2012). Higher-order infinitesimal robustness. *Journal of the American Statististical Association* **107**: 1546–1557.

Venables, W.N. and Ripley, B.D. (2002). *Modern Applied Statistics with S*, 4e. New York: Springer.

Voit, J. (2005). *The Statistical Mechanics of Financial Markets*, 3e. Berlin: Springer.

Wald, A. (1949). A note on the consistency of maximum likelihood estimation. *Annals of Mathematical Statistics* **20**: 75–88.

Wasserstein, R.L. and Lazar, N.A. (2016). The ASA's statement on p-values: context, process and purpose. *The American Statistician* **70**: 129–133.

White, H. (1982). Maximum likelihood estimation of misspecified models. *Econometrica* **50**: 1–25.

Wilcox, R.R. and Serang, S. (2017). Hypothesis testing, p values, confidence intervals, measures of effect size, and Bayesian methods in light of modern robust techniques. *Educational and Psychological Measurement* **77**: 673–689.

Wolfowitz, J. (1953). Estimation by the minimum distance method. *Annals of the Institute of Statistical Mathematics* **5**: 9–23.

Wolfowitz, J. (1954). Estimation by the minimum distance method in nonparametric difference equations. *Annals of Mathematical Statistics* **25**: 203–217.

Wolfowitz, J. (1957). The minimum distance method. *Annals of Mathematical Statistics* **28**: 75–88.

Woodward, W.A., Parr, W.C., Schucany, W.R., and Lindsey, H. (1984). A comparison of minimum distance and maximum likelihood estimation of a mixture proportion. *Journal of the American Statistical Association* **79**: 590–598.

Yang, S.S. (1985). On bootstrapping a class of differentiable statistical functionals with applications to L- and M-estimators. *Statistica Neerlandica* **39**: 375–385.

Yohai, V.J. (1987). High breakdown-point and high efficiency robust estimates for regression. *The Annals of Statistics* **15**: 642–656.

Yohai, V.J. and Maronna, R.A. (1976). Location estimators based on linear combinations of modified order statistics. *Communications in Statistics: Theory and Methods* **5**: 481–486.

Yohai, V.J. and Maronna, R.A. (1979). Asymptotic behaviour of M-estimators for the linear model. *The Annals of Statistics* **7**: 258–268.

Yohai, V.J. and Zamar, R.H. (1988). High breakdown-point estimates of regression by means of the minimization of an efficient scale. *Journal of the American Statistical Association* **83**: 406–413.

Yohai, V.J., Stahel, W.A., and Zamar, R.H. (1991). A procedure for robust estimation and inference in linear regression. In: *Directions in Robust Statistics and Diagnostics Part II* (ed. W.A. Stahel and S.W. Weisberg), 365–374. New York: Springer-Verlag.

INDEX

Robustness Theory and Application, First Edition. Brenton R. Clarke.
© 2018 John Wiley & Sons, Inc. Published 2018 by John Wiley & Sons, Inc.
Companion website: www.wiley.com/go/clarke/robustnesstheoryandapplication